AF508675

Teaching and Learning Advances on Sensors for IoT

Teaching and Learning Advances on Sensors for IoT

Editor

Sergio Martin

MDPI • Basel • Beijing • Wuhan • Barcelona • Belgrade • Manchester • Tokyo • Cluj • Tianjin

Editor
Sergio Martin
National Distance Education University
Spain

Editorial Office
MDPI
St. Alban-Anlage 66
4052 Basel, Switzerland

This is a reprint of articles from the Special Issue published online in the open access journal *Sensors* (ISSN 1424-8220) (available at: https://www.mdpi.com/journal/sensors/special_issues/teach_IoT).

For citation purposes, cite each article independently as indicated on the article page online and as indicated below:

LastName, A.A.; LastName, B.B.; LastName, C.C. Article Title. *Journal Name* **Year**, *Volume Number*, Page Range.

ISBN 978-3-0365-0522-0 (Hbk)
ISBN 978-3-0365-0523-7 (PDF)

Contents

About the Editor

Sergio Martin (Associate Professor) works at UNED (National University for Distance Education, Spain). He received a Ph.D. from the Electrical and Computer Engineering Department of the Industrial Engineering School of UNED. He is a computer engineer in distributed applications and systems at the Carlos III University of Madrid. He has taught subjects related to microelectronics and digital electronics since 2007 in the Industrial Engineering School of UNED. He has participated since 2002 in national and international research projects related to mobile devices, ubiquitous computing and the Internet of Things as well as in projects related to "e-learning", virtual and remote labs and new technologies applied to distance education. He has published more than 200 papers both in international journals and conferences and received more than 25 awards during his career. He has organized several international conferences and has served as associate/guest editor of several JCR journals.

Preface to "Teaching and Learning Advances on Sensors for IoT"

Dear Colleagues,

The Internet of Things (IoT) is widely considered the next step towards a digital society where objects and people are interconnected and interact through communication networks. The IoT not only has a huge social impact, but can also support the employability and boost the competitiveness of companies. It is widely considered one of the most important key drivers for the implementation of so-called Industry 4.0 and for the digital transformation of companies.

Sensing is a fundamental part of IoT environments, where massive amounts of data are constantly sensed. Proper quality data acquisition leads to more accurate decision making. Thus, the importance of good practices when sensing data in IoT environments is growing.

The rapid diffusion of IoT technologies has created an important educational challenge, namely, the need to train a large number of professionals able to design and manage a fast evolving and complex ecosystem. Thus, an important research effort is being carried out in innovative technologies (simulators, virtual and remote labs, mobile apps, robotics, e-learning platforms, learning analytics, etc.) applied to innovative teaching practices.

This book focuses on all the technologies involved in improving the teaching and learning process of some of the sensor-based IoT topics, such as virtual sensors, simulated data acquisition, virtual and remote labs for IoT sensing and innovative teaching materials, among others.

For example, the work presented by Pastor-Vargas et al., Labs of Things at UNED (LoT@UNED), provides remote laboratories for full IoT development, including edge, fog and cloud computing, complemented with communication protocols and cybersecurity. The use of these remote laboratories allows students to acquire complete IoT skills using real devices and platforms from home. The paper also introduces its use in an official master's degree in computer engineering. The work done by Fernández-Camarés et al. provides an introductory practical guide to IoT cybersecurity assessment and exploitation. On the other hand, Huertas et al. propose a novel multimodal learning analytics architecture that builds on software-defined networks and network function virtualization principles. The provided findings and the proposed architecture can be useful for other researchers in the area of MMLA and educational technologies envisioning the future of smart classrooms. The work done by Tabuenca et al. addresses the educational need to train students on how to design complex sensor-based IoT ecosystems. Hence, a project-based learning approach is followed to explore multidisciplinary learning processes implementing IoT systems that varied in the sensors, actuators, microcontrollers, plants, soils and irrigation systems they used. Finally, the work done by Ruiz-Rube presents several extensions to the block-based programming language used in App Inventor to make the creation of mobile apps for smart learning experiences less challenging. Such apps are used to process and graphically represent data streams from sensors by applying map-reduce operations.

The articles published in this book present only some of the most important topics about IoT learning and teaching. However, the selected papers offer significant studies and promising environments.

Sergio Martin
Editor

Article

A WoT Platform for Supporting Full-Cycle IoT Solutions from Edge to Cloud Infrastructures: A Practical Case

Rafael Pastor-Vargas [1,*], Llanos Tobarra [1], Antonio Robles-Gómez [1], Sergio Martin [2], Roberto Hernández [1] and Jesús Cano [1,3]

[1] Department of Control and Communication Systems, Computer Science Engineering Faculty, Spanish National University for Distance Education (UNED), 28040 Madrid, Spain; llanos@scc.uned.es (L.T.); arobles@scc.uned.es (A.R.-G.); roberto@scc.uned.es (R.H.); jesus.cano@computer.org (J.C.)
[2] Electrical and Computer Department, Industrial Engineering School, Spanish National University for Distance Education (UNED), 28040 Madrid, Spain; smartin@ieec.uned.es
[3] Faculty of Law, San Pablo CEU University, 28003 Madrid, Spain
* Correspondence: rpastor@scc.uned.es

Received: 22 May 2020; Accepted: 2 July 2020; Published: 5 July 2020

Abstract: Internet of Things (IoT) learning involves the acquisition of transversal skills ranging from the development based on IoT devices and sensors (edge computing) to the connection of the devices themselves to management environments that allow the storage and processing (cloud computing) of data generated by sensors. The usual development cycle for IoT applications consists of the following three stages: stage 1 corresponds to the description of the devices and basic interaction with sensors. In stage 2, data acquired by the devices/sensors are employed by communication models from the origin edge to the management middleware in the cloud. Finally, stage 3 focuses on processing and presentation models. These models present the most relevant indicators for IoT devices and sensors. Students must acquire all the necessary skills and abilities to understand and develop these types of applications, so lecturers need an infrastructure to enable the learning of development of full IoT applications. A Web of Things (WoT) platform named Labs of Things at UNED (LoT@UNED) has been used for this goal. This paper shows the fundamentals and features of this infrastructure, and how the different phases of the full development cycle of solutions in IoT environments are implemented using LoT@UNED. The proposed system has been tested in several computer science subjects. Students can perform remote experimentation with a collaborative WoT learning environment in the cloud, including the possibility to analyze the generated data by IoT sensors.

Keywords: web of things; IoT learning; cloud computing; protocols; virtualization; instructional design

---

1. Introduction

Internet of Things (IoT) [1] has become a key technology for the interconnection of smart devices [2] with their surroundings. These devices acquire information from their immediate environment using specific sensors and change the state of their environment through actuators. These changes are performed through algorithms that determine the interaction with the environment. This computational capacity is defined by the "Edge Computing" paradigm, which encompasses not only algorithmic solutions but also the boundary conditions that must be taken into account when implementing the device's intelligence [3–5]. These conditions include requirements in terms of response time, cost and energy consumption and use of bandwidth in communications, among others.

In the field of education, these technologies have been employed in computer science courses [6], by allowing students to have a smooth and natural approach to them and their applications [7,8].

Additionally, [9,10] present the evolution of IoT learning scenarios in contexts like distributed computing and cybersecurity. These contexts use distance learning/teaching methodologies and corresponding environments.

The use of IoT applications has multiple fields of application [1], such as e-Health (health monitoring of people [11], Personalized Healthcare [12] or biosensors-based environments [13,14]), Smart Cities (traffic control [15] or intelligent transport systems [16]), Agriculture [17,18] or the vehicle industry [19,20], among many others. The applications are practically endless, considering that the number of intelligent devices and sensing systems are growing at a dizzying pace.

IoT has exploded in recent years, and it does not look like a short-term slowdown is taking place. Gartner [21] predicted that there will be 20.4 billion smart devices connected and in use worldwide by 2020, and a new Business Insider Intelligence study [22] predicts that the IoT market will grow by more than $3 billion a year by 2026.

Taking into account the need for professionals in all the areas mentioned above, it is necessary to have specific learning processes that allow students to acquire the necessary competences and skills to undertake projects based on IoT infrastructures. Students must use components and layers (hardware/software) that are deployed in this type of solution, so the learning process must incorporate the use of technological tools similar to those that will be found on these IoT environments and domains. Thus, the objectives of this paper are the following:

1. Analyze the main stages involved in the IoT development cycle and define the essential characteristics of an environment that supports the learning and experimentation of all these stages.
2. Describe the main features of a system designed by the authors that cover all IoT development stages and and how this system fulfils the essential characteristics mentioned before.
3. Evaluate the students' perception of the platform's usefulness and its applicability in the different stages involved in IoT projects.

The developed platform, Labs of Things at UNED (LoT@UNED), provides remote laboratories for full IoT development, including edge, fog and cloud computing and complemented with communication protocols and cybersecurity. The use of these remote laboratories allows students to acquire complete IoT skills using real devices and platforms from home. The paper also introduces its use in an official master degree in Computer Engineering.

Regarding the paper organization, Section 2 shows the methodology followed in this paper. Section 3 describes the state of the art found in the literature about IoT remote laboratories. Section 4 describes the platform proposed by the authors, from the hardware, software and communications point of view. Section 5 describes the practices implemented with this platform in a real use case. Section 6 provides the results of a satisfaction survey provided to students. Section 7 details the discussion of the main findings from the survey. Finally, conclusions are given in Section 8.

2. Methods

The methodology followed in this study includes the following steps:

1. Analysis of IoT applications to determine the main stages of IoT development.
2. Analysis of the literature to identify previous papers published describing IoT remote laboratories. This analysis consists of the search for articles in the main scientific repositories for this topic: MDPI, IEEExplorer and ScienceDirect. This step analyzes in which stages of IoT development are focused the found papers.
3. Analysis of the system proposed by the authors to check the stages of IoT development covered.
4. In-depth description of the proposed system by the authors from the hardware, communications and software points of view.
5. Description of the experimentation of this platform in a real Computer Engineering subject, including the full IoT development cycle and indicating the designed practices provided to the

students. A methodology based on a typical flow of the instructional design [23] has been used. The first assumption made to start the application of instructional design is that the setting up of a laboratory is not an isolated task. It should be integrated into the subject objectives as another element of the instructional design. It should be a way to acquire a competence or skill related to the subject. In addition to this, the methodology involved in a teaching/learning process on distance, as our case is, implies much more periodical virtual attendance and interaction with/among students than with a traditional methodology. All these learning resources presented to students have to be available to them all time, and they expect innovation teaching approaches to improve the quality of courses. This expectation is more noticeable in Engineering subjects than only theoretical subjects such as maths when practical skills have to be acquired by students. Additionally, supporting many students becomes a real challenge when technologies are implemented and deployed in virtual courses. The instructional design methodology is made up of the four phases, as it can be observed in the Figure 1:

- *Activity Description.* The educational objectives are first defined, a global description of the laboratory is given to students and the expected outcomes are detailed to them in this step.
- *Activity Design.* In this phase, a set of elements to be employed in the activity are selected, the acquisition data mechanism and the way in which the elements interact.
- *Activity Development.* Once the laboratory design is finished, students will be required to perform some programming task with a set of provisioned services and obtained data, as well as running a set of client applications. These have to be used, tested and synchronized among them.
- *Experimentation.* The last step is to do experimentation with the services and applications deployed to make improvements.

6. Survey preparation to analyze students' satisfaction. It includes questions about: gender, age and occupation. It also includes five-point Liker-scale questions about perceived usefulness, ease of use, user attitude, social influence, ease of access and intention of use.
7. Analysis of the satisfaction survey provided to the students to validate the tool from a satisfaction point of view. The previously mentioned indicators are analyzed by studying their standardized mean, standard deviation, variance, minimum and maximum values, median, kurtosis, asymmetry and Cronbach's alpha.

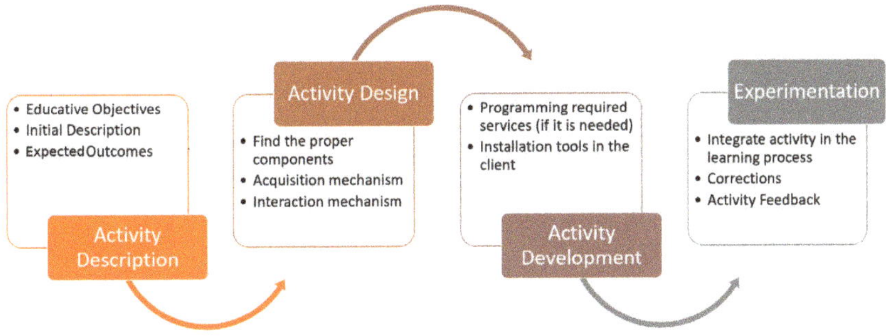

Figure 1. Instructional design phases for a new laboratory.

3. State of the Art

To understand the complexity associated with the development of IoT solutions, it is important to understand the organization of these systems, usually in a set of layers that implement specific functionalities [3,24], as it can be observed in Figure 2. Usually, these layers are classified using a criterion of physical proximity to the environment and processing capacity of the components that integrate it [25]:

- Layer 1: Edge computing [26,27]. This layer integrates hardware and software components as smart devices, sensors and IoT protocols.
- Layer 2: Fog computing [4,5]. This intermediate layer provides an extra resources layer, such as computing power or real-time services, to the edge layer.
- Layer 3: Cloud layer or dashboards [28] and assisted decision layer [24,29]. Data based solutions using cloud services and the storage service of sensor data in layer 2. Usually, this layer uses services, which need a high computational power capacity so this layer can be integrated with the cloud provider of layer 2 or be located in another cloud provider, such as AWS IoT [30], Microsoft Azure IoT [31] and IBM Watson IoT [32], or specific platforms [33].

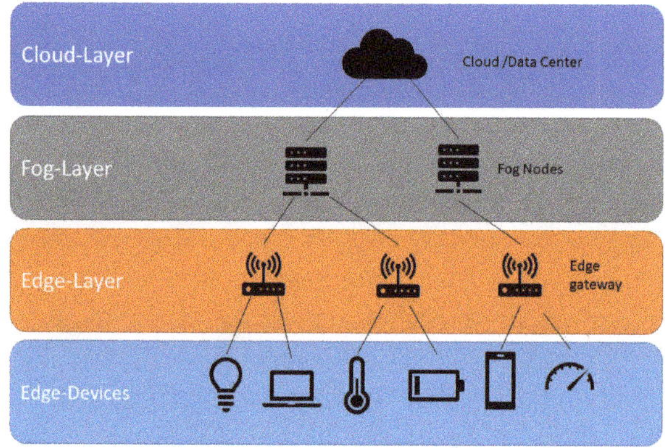

Figure 2. Fog computing approximation for Internet of Things (IoT) solutions. Figure available on [34].

Other important aspects to have into account when analyzing IoT applications are the communication technologies and protocols (such as HTTP, MQTT [25], CoAP [35] and others [36–38]), and cybersecurity.

Analyzing the literature, most of the educational IoT labs are for hands-on experimentation. Among those designed to experiment online out of laboratory facilities, many were just pure simulations or virtual labs [39]. An example is the work of Patil et al. [40], who describe an IoT virtual lab to allow students sensing and retrieving simulated data from the cloud using Python, as part of the modeling and simulation lab.

Only a few remote labs can be found to allow remote experimentation. An example is the work of Tunc et al. [41], who presented an IoT remote laboratory designed only for cybersecurity experimentation.

El-Hasan [42] introduces an IoT mobile dashboard to allow off-campus practices through a system including sensors, controlling and interfacing kits, cameras and others. Basically it only allows the modification of certain parameters to switch the direction of rotation of a motor by changing predefined values of voltage, current and power as well as other required parameters, such as speed and torque.

Fernandez-Pacheco et al. [43] describe an Arduino remote lab using a Raspberry Pi as a server, but it is only intended for microcontroller programming (Arduino), not for IoT purposes (cloud, Python programming, IoT protocols, cybersecurity, etc.).

Leisenberg [44] presents a remote lab based on Raspberry Pi for movement analysis. Students should write the code to analyze real time images coming from a webcam. Again this system is not intended for full-cycle IoT purposes.

Rajurikar et al. [45] present a system for IoT protocols experimentation, mainly REST and CoAP. They connected several sensors through an Arduino board and a Beagle Bone device to a cloud platform. This cloud environment is only intended for End point Data Acquisition and decision-making.

From the previous work, we can conclude the essential characteristics to be covered in a IoT laboratory/environment are:

- Access and development with IoT devices (edge programming).
- Development of solutions in the fog, with limited computational capacities (fog programming).
- Analysis of the data provided by the sensors and interaction with the actuators of the device infrastructure (cloud dashboard and analytics programming).
- Configuration and management of specific communication protocols for IoT (protocol experimentation).
- Development of security techniques in IoT environments (cybersecurity).

It can also be observed that none of the analyzed works allow the implementation of all the essential features required for learning the complete development cycle of the IoT applications. Only Rajurikar et al. [45] complete the implementation and support of three out of the five features. As a consequence, it is necessary to have an environment that complies with all the characteristics.

Table 1 compares the functionality implemented in each of the IoT remote labs found on the literature and the authors' proposed system. It can be observed that our proposal covers all the features included in the study (edge programming, fog programming, cloud dashboards and analytics programming, protocol experimentation and cybersecurity), whereas other approaches only cover one or several functionalities. This way, the development of LoT@UNED implements the full set of features, advancing in the development/research of this type of environments.

Table 1. Functionality comparison of the state of the art on IoT remote labs.

	Edge Programming	Fog Programming	Cloud Dashboard and Analytics Programming	Protocol Experimentation	Cybersecurity
Tunc [41]					X
El-Hasan [42]	X				
Fernandez-Pacheco [43]	X				
Leisenberg [44]		X			
Rajurikar [45]	X		X	X	
Authors	X	X	X	X	X

To check how these features are implemented in the authors' proposed system, the following section describes the LoT@UNED platform more in detail.

4. Solution Description

4.1. Hardware Architecture

The LoT@UNED platform implements the edge layer through a set of IoT devices (i.e., Raspberry Pi boards). Each device is connected to the services of layer 2 (Cloud IoT Layer) using the MQTT protocol. This way, students can develop the skills and abilities corresponding to layers 1 and 2. The services of the Cloud IoT layer are provided through the IBM cloud provider, and specifically using the IBM Watson IoT service. The service for storing sensor/device data is also implemented in this provider. The non-relational database Cloudant is used for this purpose [46]. This cloud storage service is used in the dashboard and assisted decision layer (Cloud Layer). Again, the IBM Watson Studio service from IBM Cloud is used for the development of the analysis and machine learning algorithms based on the data stored in the Cloudant service. LoT@UNED has been designed flexibly to be able to use other specific services from other providers for the cloud layers, so that experiences can be designed for the development of IoT solutions using the AWS or Google services (AWS IoT, Cloud IoT, S3 or Cloud Storage). The basis of a previous generation of our Web of Things (WoT) platform was presented in [10].

The structure of the platform focuses on the availability of low-cost Raspberry Pi devices, which are connected through a cluster that eases the device connection to the Internet. There are two logical groupings available that facilitate the connection of new devices. The first group uses a specific rack for the management and connection of the devices, as shown in Figure 3a. The characteristics of the rack can be found on the website of the BitScope provider [47]. It allows grouping up to 40 Raspberry Pi (Model B) devices, facilitating the connection to the electricity grid, the connection among devices and the Internet connection. It can also be installed in a traditional rack, facilitating the management of the cluster itself. Its high cost and the inability to add specific sensors in an individual way for each device can be noticed as its main drawbacks.

The second grouping, as shown in Figure 3b, uses cheaper and more flexible components in terms of device separation. This fact allows us to add specific sensors (cameras, GPS, temperature sensors, etc.) without storage problems or cluster connectivity. In fact, the storage can be increased by lateral fixings that support the structure of the cluster. Proof of this is the specific configuration that is used in the example specified in the following section. This type of configuration is deployed outside of the clusters to ease the replacement and management of the sensors in order to provide the necessary redundancy for the services that uses this configuration.

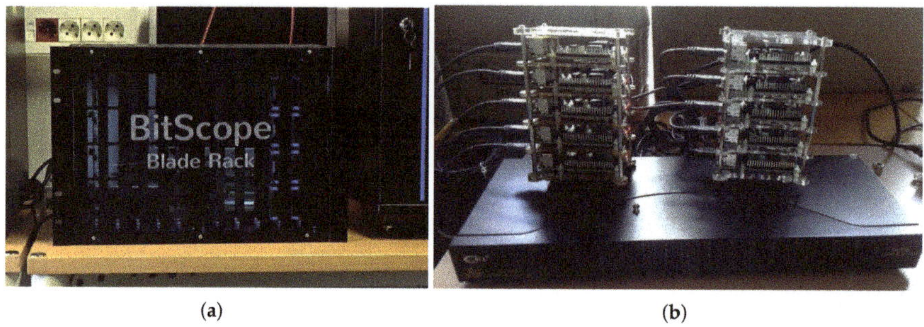

(**a**) (**b**)

Figure 3. Devices clustering: blade rack and cheap setup. (**a**) Blade rack. (**b**) Cheap setup.

The variety of setups (clustered or individual) allows the logical grouping of the services offered by IoT@UNED and, also, the redundancy necessary to provide a stable learning service for students. For example, in the case of the setup of Figure 5, there are three exact replicas that are managed by the software developed, installed on the base image of the Raspberry Pi and integrated transparently within the service availability (a service, three concurrent accesses). The base image of each Raspberry Pi card comes with the connection services to the IoT service in the cloud, which allows the self-registration of the devices. This self-registration allows us to automatically have the inventory of the devices and the available setups for the entire IoT environment of IoT@UNED. Each device will add specific information about the type of educative service offered (it can be more than one), which will allow the activity manager to decide on the assignment of each environment for the student who requests it.

4.2. Communication

Resources in IoT@UNED are understood as a standard communication channel using the MQTT protocol (for interaction commands) and the required software to "control" and "program" the device (a Python distribution, sensor access libraries, etc.). All these resources define a run-time environment that depends on the service that the end-user wants to offer.

The MQTT protocol has been selected due it its popularity and specific way of working. This paradigm uses the message as a fundamental unit of communication. The participants in the solution are the ones who give meaning to the message. The roles of subscriber, editor (publisher) and broker are usually assigned. Subscribers register their interest in certain messages from one or more editors/publishers. This interest is handled through the broker, which is the responsible for managing the flow of messages between publishers and subscribers. Once the editor generates a message, it is delivered through the broker so that it sends it to the interested subscribers. Hence, the MQTT protocol is able to simplify and facilitate the synchronization between all nodes and jobs available in the IoT platform.

Each message with MQTT is associated with a topic, so the broker and the subscriber can identify the message. The usual topics are "data", "status" or "alarm"; and they act as semantic labels of the information carried by the messages. The targeting of different topics allows administrator to check the health of the IoT solution and to monitor its network communication in a fast way.

An example of a message's flow for MQTT is shown in Figure 4. In this particular case, a temperature sensor (publisher) is sending the temperature using the topic data to the MQTT broker. The MQTT broker delivers the topic messages to the two subscribers (computer and mobile device), which previously registered their interest in the data topic.

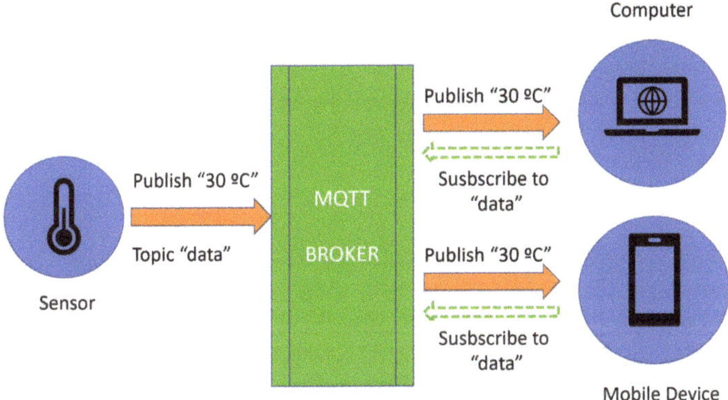

Figure 4. MQTT flow of messages for topics and subscriptions.

4.3. Software Architecture

4.3.1. Virtualization and Orchestration

The run-time environment can be "packaged" using already known virtualization technologies [41], such as Docker [48,49]. Docker is based on the use of containers that define a prefabricated execution environment. Docker can be deployed in any infrastructure that supports this technology. The definition of a service is based on the execution of one or more Docker containers, although usually only one of them is necessary. Specifically, in the case of services associated with experimental sessions with IoT devices, the container is executed on the same device which provides sensors and runtime. However, in more advanced practices it is possible to run several containers on the same device or several at the same time. This orchestration of containers allows identifying scenarios of collaborative use where the sensors of several devices [50,51] are used in coordination to obtain a specific purpose (traffic control at crossings with several traffic lights, data from the environmental sensors of several drones flying over aerial areas for pollution indices, etc.).

Regardless of whether the service requires the execution of one or more containers, it is essential to provide an orchestration layer of those containers providing:

- Dynamic access management to devices. Since there may be an inherent concurrence in the development of practices in a remote environment, the orchestration layer must identify which services/containers are running on the IoT devices. This capability facilitates the search for containers/devices available in the LoT@UNED infrastructure and the assignment to new students.
- Redundancy and fault tolerance. The orchestration layer identifies the number of IoT devices per usage scenario (service). It is able to assign, in case of failure, a new device (or several, depending on the service). The ability to re-start (resume) the work session in case of failure is not currently supported.
- Management of the basic containers of the services. To facilitate the distribution of existing containers/services or the distribution of new ones, the orchestration layer should have the ability to locate the images of those containers in standard repositories [52].

There are several orchestration system solutions available, such as Docker Compose [53], Docker Swarm [54] or Kubernetes [55]. For the orchestration and control layer of LoT@UNED, Kubernetes has been selected since it eases the management of the device containers and the supervision of all the executed containers in the infrastructure through its dashboard [56]. This characteristic is essential to provide a continuous service delivery of the IoT laboratories in the infrastructure and an availability close to 24×7.

The main drawback of the Kubernetes deployment model is associated with the dedicated use of one of the infrastructure devices as the master node of the orchestration layer. This makes the orchestration layer vulnerable to the fall of this node and, therefore, it is essential to monitor it in real time, and to include automatic restart procedures.

4.3.2. Execution Services

The complete architecture of the execution services into the LoT@UNED infrastructure is shown in Figure 5. It shows how each IoT device contains a Docker run-time environment and acts as a slave node of the Kubernetes cluster. On the control plane, there is a device (a Raspberry Pi) that acts as a master of the cluster and it communicates with the broker (IBM Watson IoT) to ease the communication channel (MQTT) with the IoT devices to be used during the interactive sessions (Shell Service).

There are currently three base containers that are identified with the "services" offered by the LoT infrastructure:

1. *IoT*. The container provides a run-time environment based on the Python programming language and the sensor access libraries are available in device setups with the Sense Hat module. It is used in the field of knowledge of IoT solutions.
2. *Programming*. This container/service only provides an environment with a Python distribution for its use in basic programming activities.
3. *Security*. This environment provides the basic Linux tools for cybersecurity operations through a virtual shell console: nmap, wireshark, route, etc. commands do not really run on the provided virtual console, but directly on Raspberry Pi 3 devices, through the service orchestration platform.

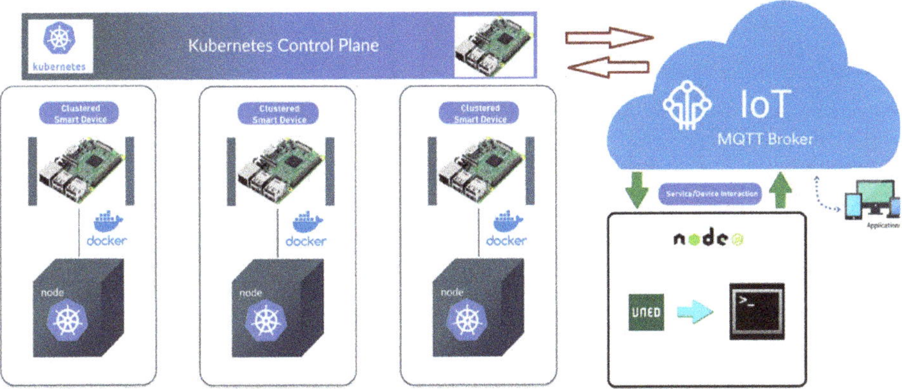

Figure 5. Technical solution for Labs of Things at UNED (LoT@UNED).

With these computing services (of execution) it is possible to build and define laboratory activities in different areas using the flexible infrastructure of LoT@UNED. The fundamentals of the definition of learning services and the workload protocol to define them using the tools/applications provided by LoT@UNED are detailed in the next subsection.

4.3.3. Services Implemented

In order to take advantage of the scalability of the infrastructure introduced above, it is necessary to provide such infrastructure with a set of services that allows the use of devices in remote educational environments. The offered services should implement the following features:

- Authenticate the user (student/teacher) into the infrastructure. It is important to facilitate the automatic identification of users. So, users must log in once but be able to access all the services transparently (SSO, Single Sign-On).
- Provide direct access to an interactive environment with devices. This environment is customized for the practice that the student must perform. Therefore, the actions that students can perform on the devices are limited by the environment configuration.
- Include analytic capabilities by storing the student's interaction through the whole cycle with the devices. Thus, the executed commands as well as the responses can be retrieved for review. Consequently, lecturers can evaluate the student's performance during the work session.
- Provide the capability to create and edit learning practices using predefined services for a specific field of knowledge (for example, IoT).

These characteristics are implemented through a set of services and applications that are included in the environment. Specifically, two fundamental applications are fully integrated with LoT@UNED:

- *Initial web portal for students/teachers* [57]. This website portal allows user authentication and access to the different practices available to the student, grouped by area of knowledge (see Figure 6). In addition, in the case of the teacher role, practices can be created from predefined execution services, adding the appropriate learning resources (statement of practice). Under the teacher role, work session options can also be configured (duration, commands to be executed on the IoT device, etc.). The access portal also allows verifying and analyzing the work sessions in the different practices, intending to evaluate the students.
- *Shell.* This application implements direct interaction with the IoT device, taking into account the possible actions and configuration of the practice defined by the teacher (Figure 7). The MQTT protocol is used for interaction with the devices, which allows the entire work session to be stored.

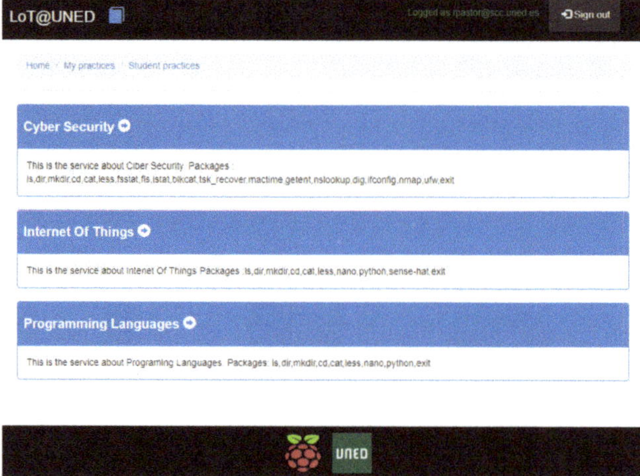

Figure 6. Knowledge domain for IoT supported learning scenarios.

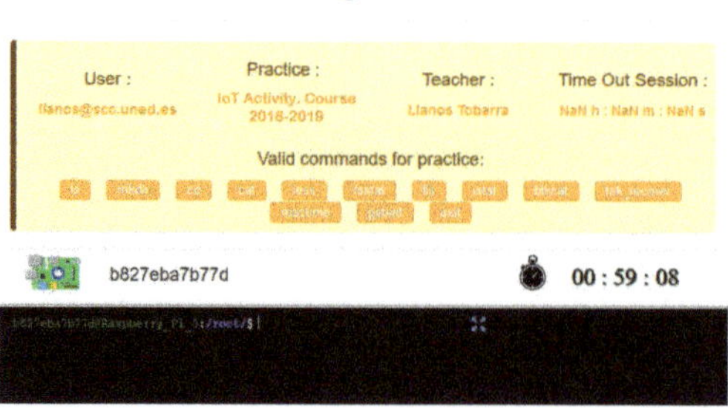

Figure 7. Interaction with IoT devices using the Shell service in LoT@UNED.

5. Experimentation

The sample case is carried out within the context of the "Cloud Computing and Network Service Management" subject, which belongs to the MSc degree in Computer Science Engineering. This degree is composed of a set of mandatory and optional subjects, some of them having 6 ECTS credits and others 4 ECTS credits. The subject considered in this work is mandatory, consists of 4 ECTS credits and is studied in the first semester of the first academic year. The degree is taught at the Computer Science Engineering School of the public Spanish University for Distance Education (*in Spanish, Universidad Nacional de Educación a Distancia*, UNED). The learning/teaching methodology is totally on distance, since Master degrees at UNED do not consist of face-to-face classes.

The subject focuses on specific competencies and skills in developing cloud computing solutions [58]. Students are provided with a guided example on the use of these technologies over a complete IoT solution. Three different and interconnected practical activities have to be solved by the students of this subject. The cybersecurity practice was not used, as it was out of the subject syllabus:

1. Development of a simple application in a cloud service provider.
2. Connecting IoT devices to IoT Cloud Services and Platform from a cloud provider. These IoT devices are real boards accessed remotely.
3. Development of dashboards and data analytics based on information provided by sensors connected to IoT devices.

This solution is especially important for distance educational environments, which should satisfy the following requirements:

- 24 × 7 availability of services associated with smart devices.
- Dynamic management of smart devices (horizontal growth of the IoT solution).
- Integration with IoT platforms and services in the cloud through the use of standard communication protocols, such as those mentioned in the previous section.
- Direct interaction with the sensors/elements of the devices through simple user interfaces and using Web protocols.

The following sections describe each one of the practices implemented.

5.1. Practice 1: Simple Application in a Cloud Service Provider

As it was mentioned earlier, the use of IoT devices is required in the second assessment. In this case, the specific skills to be learned focus on the first layer of the development of IoT solutions, this is, the sensors/devices/protocols layer. In this practice, students have to deal with three of the essential characteristics: edge programming, fog programming and protocol communication. A specific setup is integrated into the LoT@UNED infrastructure for providing students with remote access to this working layer (IoT devices). This setup is replicated, and it consists of a Raspberry Pi device and its corresponding sensors. These setups are connected to the LoT@UNED infrastructure, so they are available to the students by using the service portal. Each IoT device is able to record video and capture photos, as well as measure temperature, humidity and pressure. It also captures values associated with motion/location sensors (gyroscope, accelerometer and magnetometer), and it includes a GPS module in anticipation of future mobile scenarios.

The physical implementation of this setup is carried out with a Raspberry Pi 3, as the basis of the device/microcontroller component. This device, by default, does not have any specific sensor/actuator, but many of them can be connected to develop different projects. In this specific case, a set of additional elements has been incorporated to generate an environment with a set of sensors. These elements are:

- Raspberry Pi Camera. This element provides the features of video recording and photo capture. It can also be used in remote space surveillance projects, configuring the device to broadcast in real-time streaming.
- USB microphone. Since the Raspberry's operating system is Debian, it is possible to connect standard devices to its ports (specifically to the USB ports). In that case, the audio recording has been added to the device to complement the video recording.
- GPS module. Although our learning scenario is considered to be static by default, this module has been added in anticipation of mobile scenarios. In addition to this, it provides with a very rich dataset in terms of GPS position itself, measurement error data and other associated values provided by satellites when reading GPS values.
- Sense Hat module. This module was originally created to work on the Astro Pi mission in the international space station. Subsequently, it became widely available to the entire Raspberry user community. The Sense Hat module (see Figure 8) provides temperature, humidity and pressure readings, as well as the values associated with motion/location sensors (gyro, accelerometer and magnetometer). Additionally, it provides an array of 8 × 8 LEDs (RGB) and a five-button joystick.

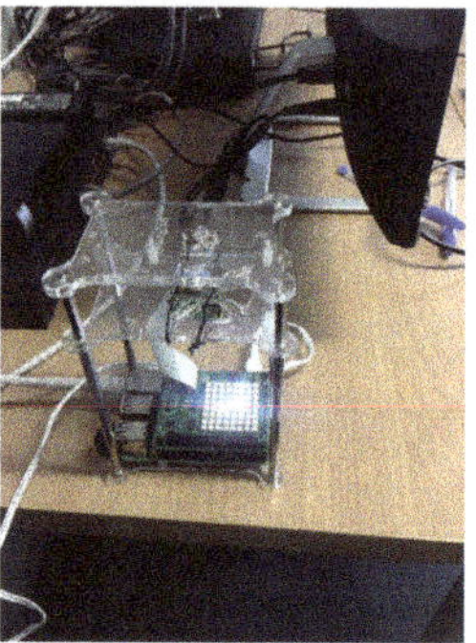

Figure 8. Example of individual Raspberry Pi device.

Students can access the Shell console to interact with the sensors by developing code in Python. This code is used to get sensor values (temperature, pressure, humidity, accelerometer data and GPS data). They can also write values directly on the LED array, so a word or phrase can be displayed in the array. The setup provides a video stream that can be programmed using python code (starting and stopping the video stream). To implement this activity, the related practice is designed using the corresponding runtime service "IoT". This service is defined as one of the three runtime services available, as it was mentioned earlier.

The service is configured to connect with a Cloud IoT Service Platform (IBM Watson IoT). This way, the setup's environment provides the MQTT library (owned by IBM and deployed on the setup) for programming and implementing the MQTT services (messages, topics and so on). These services must be deployed via Python code and consumed by an external application, which has to be developed by students (similar to the application shown in Figure 9).

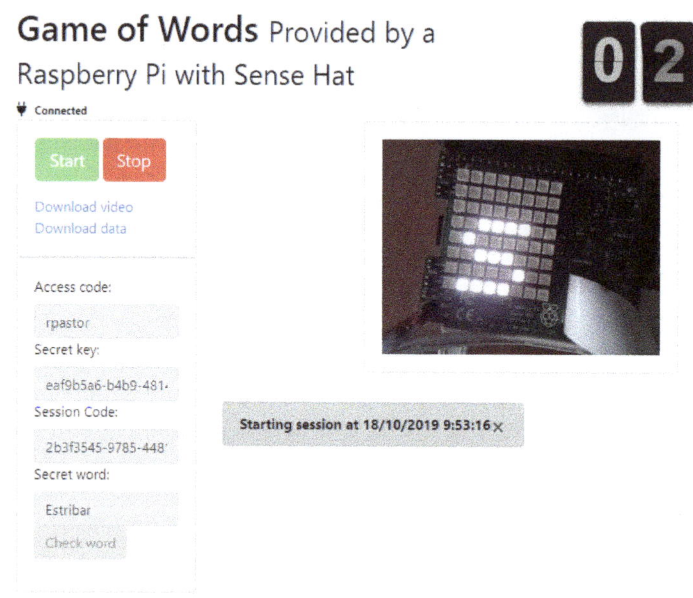

Figure 9. Game of words. IoT devices with sense hat.

Node-RED [59] is used to program this application. This framework is a visual development tool for programming IoT environments. As the setup uses MQTT, students have to use connectivity blocks for sending and receiving MQTT messages or topics. On the one hand, this tool allows the subscription to one or several topics of the MQTT message system (coming from one device, or several ones); such as the data from the sensors, current sessions or stops and messages between them. On the other hand, a topic (message or event) to the MQTT message system for session management can be published.

5.2. Practice 2: Connecting IoT Devices to IoT Cloud Services and Platform from a Cloud Provider

After developing the "local" solution, corresponding to the sensors/devices/protocols layer, students must learn and know the operation and services of an IoT platform in the cloud. In this case, it is specifically intended that they learn how to store data from the sensors they are using in the local solution. Since MQTT is used as a communication protocol, any cloud service platform that supports this protocol can be used in this part of the learning scenario (layer 2 of the IoT's full development model). The platform used by the students for their practices is IBM Watson IoT because the LoT@UNED infrastructure itself is based on this platform.

The main objective of this activity is to become familiar with the use of a series of services offered by IBM Watson IoT, focusing on the storing of sensor data and device management. IBM Watson IoT has a management space for device types and registered devices. Again, to understand the services provided by the Cloud IoT Platform, a student must use a specific activity defined in LoT@UNED. This practice is based in the "IoT" runtime service and its goal is to connect with the management space (using MQTT protocol) and check the services for this Cloud IoT Platform. The full documentation and services description is available in [60].

Additionally, to provide a cloud storage service, students must develop a single cloud application, which uses the Cloudant [61] service to store the sensor 's data. This application uses Node-red framework to facilitate the integration with the MQTT protocol and get the data from the device (assigned using LoT@UNED infrastructure). The Node-red distribution, included as a service in the IBM Cloud platform, has specific blocks to connect with Cloudant services to simplify the storing of information (see Figure 10). This data will be used in the next step of the learning scenario for the Layer 3 of our sample case.

In short, the practice focuses on the aspects related to the specific communication protocols of IoT and the integration with external suppliers. In addition, students experiment with the security mechanisms of these protocols and the applications/services that use them. For the essential characteristics, students work on: protocol experimentation and cybersecurity.

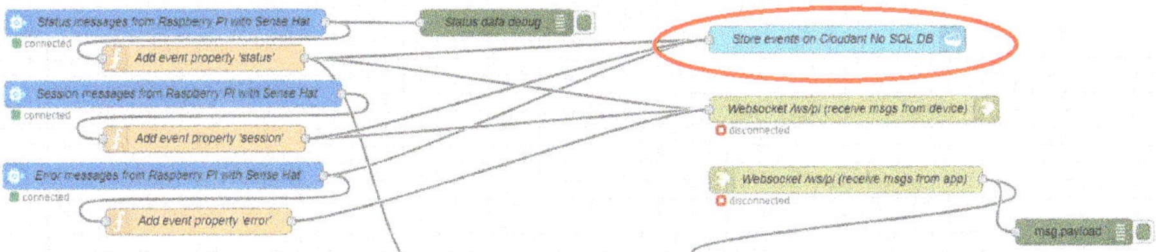

Figure 10. Cloudant integration in a Node-Red application.

5.3. Practice 3: Development of Dashboards and Data Analytics

In previous sections, the theoretical aspects of dashboards and assisted decision layer learning were given and, also, how this layer is linked with the LoT@UNED platform. Now, we detail a concrete example of an application for this layer.

The presentation and decision layer provides human-readable information to see what is the status of the IoT solution (specifically, status and information data from sensors). Sensors produce valuable information from the environment in which they are integrated. This information allows the generation of indicators to monitor different types of environments where sensorization can be critical. For example, in the case of medical environments, biomedical sensors allow information to be collected and displayed on dashboards to monitor patients [62,63]. The importance of the development of these dashboards depends on the information monitored, but usually, at least, a dashboard is developed to have monitoring information of the IoT environment. As previously seen, the information from the environment is stored in a data storage service that is usually in the cloud. This information can be represented in real-time, by dashboards, or analyzed to calculate performance indicators. These indicators can be used in decision-making and risks evaluation [46]. Therefore, these decisions are assisted by IoT data.

In this particular practice, the student will work on the development of a dashboard that uses the analytical capacity of the cloud providers and will represent the relevant information from the IoT data. This way, students will work on the essential characteristics corresponding to cloud dashboard and analytics programming. The dashboard must show real-time information about temperature, humidity and pressure (provided by the Sense Hat sensors). In addition, other indicators can be shown dynamically, such as time, the accelerometer values (X, Y and Z coordinates), pitch, yaw and roll. Figure 11 shows an example of a single dashboard built with basic gauges. This example is a basic template provided to students, which has to be modified and enriched following basic visualization techniques.

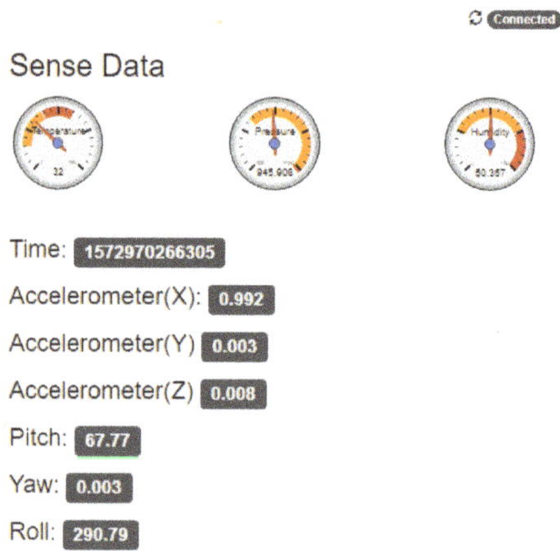

Figure 11. Single dashboard for sensor data.

Once the student has proven to be able to solve IoT data representation problems, they have to add the decision element in the production line related to an IoT solution. These decisions must be based on using stored data from sensors, and they will vary depending on the environment in which the solution is integrated [64]. The use of instantaneous data is not enough, so it is necessary to verify the evolution of the data and calculate performance indicators (usually statistical indicators). The cloud IoT service platform stores the data (layer 2 in our model), so the indicators can be analyzed and graphically represented. Data being in the cloud allows for its use in different service providers that have data analysis tools and in many cases, given the nature of the IoT data production ratio, Big Data techniques must be applied.

On the other hand, by using the Apache Spark analytics service, students have to analyze the stored data. Another service named IBM Watson Studio is used jointly with the Apache Spark Engine to facilitate the use of the analytics service. These services allow for the creation of notebooks in a variety of programming languages (among others, Python or Scala) for interactive work with the aforementioned services. Students must use one of the following sensed parameters as a basis of the analysis: temperature, humidity, or pressure. Additionally, some filter tasks are necessary for data. Some examples are changing the sensed time from string to DateTime, grouping values, filling empty values, transforming data to make specific accesses or truncating values. This way, students learn how to manage the generated data in the cloud and prepare them for the data analysis itself.

6. Results

This section analyzes the data obtained from an opinion survey provided to LoT@UNED students. Some preliminary results and conclusions were included in [65]. The amount of surveyed students was 129, in which 89.15 % of the users were male and the 10.85 % of them were female, as indicated in Table 2. With regard to the job occupation, a big amount of students are not related to computer science. In particular, a total of 79.9 % of students.

Figure 12 shows the comparison of the job situation of the surveyed students about the LoT@UNED platform, in terms of their job profile (computer scientist, non-computer scientist and others) with their age divided by ranges (less or equal to 30 years, between 30 and 39 years, between 40 and 49 years and equal to or older than 50 years). As observed, many students are in the range of 30 and 39 years old with a dominant computer science profile. The conclusion about the job occupation is even stronger for the range of 40–49 years old. In contrast, the youngest and oldest students have an occupation profile out of computer science.

Table 2. Users' profile.

Indicator	Options	(%)
Gender	Male	89.15
	Female	10.85
Age Group	$\leq$30 years	30.23
	30–39 years	42.64
	40–49 years	24.03
	$\geq$50 years	3.10
Occupation	Computer science related job position	20.1
	Non-computer science related job position	51.2
	Others	28.7

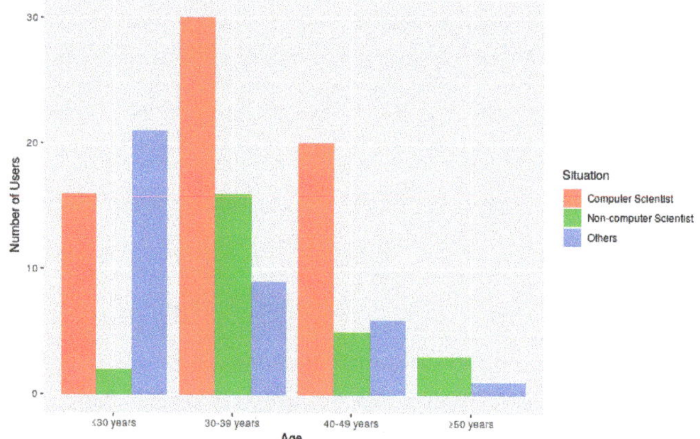

Figure 12. Comparing the job situation versus the age ranges for users who tested the LoT@UNED platform.

The measured indicators were the perceived usefulness of the LoT@UNED platform by students, its ease of use for practical activities, the users' attitude when using the platform, the social influence when using it, the ease of access to the platform and the students' intention of use the platform for practical activities within the context of LoT@UNED.

Table 3 represents the statistical data generated from the students' opinion survey (perceived usefulness, ease of use, user attitude, social influence, ease of access and intention of use), in terms of the standardized mean, standard deviation, variance, minimum and maximum values, median, kurtosis, asymmetry and Cronbach's alpha. Regarding the mean values of indicators, with a five-point scale, they can be considered as very good. The best one is the ease of use with a value of 4.13, but the worse one is the ease of access with a value of 3.40. This fact can be due to the student's profile described above. The presented standard deviation and variance values are not so high, enforcing the goodness of the exposes results. In addition to this, mean and median values are very similar. The analysis of the kurtosis, asymmetry and Cronbach's alpha indicators indicate that these results are consistent.

The kurtosis characteristic describes the concentration of data around the average of each indicator shown in Table 3. These kurtosis values are positive for four indicators (they are on the right side of the mean) and negative for two of them (they are on the left side of the mean). These characteristics consider the standardized mean of each indicator as a central point, so the data distribution is close to each indicator mean. This means they are not too scattered and in ranges of normality. This is enforced by examining the median value of each indicator, since they are near its corresponding mean.

On the other hand, the asymmetry characteristic measures the degree of symmetry of the data distribution for each indicator shown in the horizontal axis. These asymmetry values are negative in all cases, except one of them, so their distribution generally tends to the left within the x-coordinate axis. Obviously, the positive case is to the right side. They are not too high of values, so they are considered as a good distribution.

In addition to this, the Cronbach's alpha for each indicator is bounded among 0.87 and 0.90. These values are considered as more than acceptable. What is more, the general Cronbach's alpha is slightly higher than 0.9. This means that the reliability of all indicators together is really good, and we can conclude that there is a correct internal consistence. The Cronbach's alpha calculates the mean of the correlation among the exposed indicator.

These results are very rich, since they contain lower and higher values, as indicated with the minimum and maximum values. To sum up, the exposed statistical values are satisfactory, by considering the literature [66,67], being very reliable to be employed in further studies. Additionally, Table 4 shows the amount of students who answered for each indicator: strongly agree, agree, neutral, disagree or strongly disagree.

Table 3. Results obtained from an opinion survey after testing the LoT@UNED platform (statistical data).

	Perceived Usefulness	Ease of Use	User Attitude	Social Influence	Ease of Access	Intention of Use
Standardized Mean	3.93	4.13	4.11	3.67	3.40	4.04
Standard Deviation	0.87	0.95	0.89	0.79	0.71	1.04
Variance	0.76	0.91	0.80	0.63	0.50	1.09
Minimum Value	1.00	1.00	1.00	2.00	1.50	1.00
Maximum Value	5.00	5.00	5.00	5.00	5.00	5.00
Median	4.00	4.33	4.25	3.67	3.50	4.33
Kurtosis	0.96	0.15	0.43	−0.92	−0.04	0.01
Asymmetry	−0.95	−0.96	−0.93	0.21	−0.27	−0.96
Cronbach's Alpha	0.88	0.89	0.87	0.89	0.90	0.88

Table 4. Results obtained from an opinion survey after testing the LoT@UNED platform (counting with a five-point Liker-scale).

	Strongly Agree	Agree	Neutral	Disagree	Strongly Disagree
Perceived Usefulness	43	58	20	7	1
Ease of Use	61	37	21	9	1
User Attitude	61	40	21	6	1
Social Influence	22	46	55	6	0
Ease of Access	7	63	50	9	0
Intention of Use	61	34	19	13	2

Finally, Table 5 indicates how the selected indicators are correlated among them. The represented values enforce the conclusions obtained for the statistical data described above. There is a strong influence among them. The next step would be to examine their concrete influence, and how they are related. The perceived usefulness influences the user attitude about using the LoT@UNED platform in very a strong way, with a value of 0.813. The usefulness indicator also affects the intention of use of this platform in the future in an indirect way. This value is 0.689. Another strong influence is the user attitude versus the intention of use, with a value of 0.768. The rest of the indicators are very influenced among them in a lower manner. Results marked with * correspond to the presented ones in the same table when comparing the two indicators in the opposite axis.

Table 5. Correlation matrix among the exposed indicators.

	Perceived Usefulness	Ease of Use	User Attitude	Social Influence	Ease of Access	Intention of Use
Perceived Usefulness	1.000	0.597	0.813	0.615	0.552	0.689
Ease of Use	*	1.000	0.609	0.527	0.582	0.597
User Attitude	*	*	1.000	0.616	0.610	0.768
Social Influence	*	*	*	1.000	0.517	0.543
Ease of Access	*	*	*	*	1.000	0.535
Intention of Use	*	*	*	*	*	1.000

7. Discussion

According to the results obtained in the student survey, students find the LoT@UNED platform very useful (standardized mean for perceived usefulness: 3.93). It is a global indicator associated with the activities carried out in the LoT@UNED platform, described in the experimentation section. Being a non broken down indicator makes it impossible to get a particular statistical measure for every essential characteristic. However, taking into account the population associated with the survey and the statistical result, we can infer that the platform is useful for implementing each of these essential characteristics (edge, fog, cloud and analytics, protocol and cybersecurity). This inference is based on the fact that the design made for the three practices implements the five essential characteristics.

It is also interesting to note that the ease of use and intention of use have high values (standardized mean: 4.13 and 4.04, respectively). This issue means that the design of the platform itself has been done correctly and has simplified the development of the practices carried out by the students. Moreover, the high value of the intention of use indicator allows inferring that the student would be willing to use it in more similar practices in the IoT laboratory environment and even in other disciplines (cybersecurity, programming, etc.). This ability is possible because the platform allows the generation of specific services supported by the set of IoT devices that compose it.

The lowest values of the indicators (even though, they are values that indicate good behavior) correspond to ease of access and social influence indicators (standardized mean: 3.40 and 3.67, respectively). From these values, it can be deduced that the laboratory's workflow and the way to access the devices may be improved. This issue mainly affects the essential features of edge and fog computing because the virtualization layers introduce delays and complexity in the interaction that influence the ease of access and interaction. Additionally, the social influence indicator warns about the lack of interaction between students inside and outside the LoT@UNED environment. Practices are indeed carried out at distance and individually, so the social factor has less influence than in a face-to-face environment, but it is necessary to work on it. For this reason, the platform must provide collaborative tools that facilitate social interaction and communication in real time between students and teachers. These new features will be included in future versions of the platform. These improvements will focus more on the design part of learning than on the development of IoT lab environments and the support of essential characteristics for educational IoT laboratories.

8. Conclusions

The learning/teaching processes in the development cycle of IoT solutions imply a set of skills ranging from devices and IoT sensors, their communication protocols, the storage management and the processing environments on the Cloud for data generated by sensors. These environments are then eventually able to make decisions or show the relevant information on those sensors (as indicators). These fundamental competences are needed in the full cycle of development of IoT solutions, consisting on three layers: (1) basic interaction with sensors and specific communication protocols;

(2) data management models to handle the generated data; and (3) processing and visualization of the most relevant indicators on these IoT devices. In this last step, the processing can include a specific communication protocol. This protocol could be used to perform actions in the IoT device itself as a response to the processed indicators (for example, using available actuators at the device).

According to this, this work first presents the main features of the LoT@UNED platform, which has been developed to cover the instructional design of our subjects, and how the three layers of the proposed full cycle of development for IoT solutions are implemented in it. The essential characteristics for this kind of laboratories/environments are fulfilled by this platform: edge programming, fog programming, cloud dashboard and analytics programming, protocol experimentation and cybersecurity. Each phase is associated with a specific activity that is deployed in a standard way using Docker containers managed through a cluster manager (with Kubernetes). The manager balances the workload of different devices. Thus, the use of the devices/sensors is assigned in a dynamic way to the students who are developing the activities. This platform allows students to implement all these phases efficiently and redundantly, providing high availability for its use.

The proposed LoT@UNED platform has also been used for students in several computer science subjects. The use of this platform is especially relevant in online educational environments, as is the case of distance universities. This way, they perform remote experimental activities with a collaborative IoT learning infrastructure in the cloud, analyze the data generated and make visual representations in it. As for the result and discussion sections, we can conclude that the perceived usefulness and ease of use of the proposed platform values are really good, as well as the intention of use it in the future for additional practices. The students' attitude is also great with respect to the use of the platform in practical activities. The rest of the indicators are good, although they are challenging for working on improving the social influence among students when using it, and easing the access mechanisms.

As for future work, the presented method for validation of the IoT platform will be improved. To achieve this, a UTAUT model will be hypothesized. The same set of factors will be considered (easy of use, usefulness, attitude, social influence, . . .) to be included in this model, in order to check the intention to use the presented technology. Another future line of research is to exhaustively analyze the students' learning progress into the LoT@UNED platform. Finally, the source code of this tool has not yet been shared with any other institution but the release of the code is also one of our next steps for future work. We would like to have a community around it to go on including improvements.

Author Contributions: Conceptualization, R.P.-V., L.T., A.R.-G., S.M., R.H. and J.C.; Data curation, R.P.-V., A.R.-G., S.M., R.H. and J.C.; Formal analysis, R.P.-V., L.T., A.R.-G., S.M., R.H. and J.C.; Funding acquisition, R.P.-V., L.T., A.R.-G., S.M., R.H. and J.C.; Investigation, R.P.-V., L.T., A.R.-G., S.M., R.H. and J.C.; Methodology, R.P.-V., L.T., A.R.-G., S.M., R.H. and J.C.; Project administration, R.P.-V., L.T., A.R.-G., S.M., R.H. and J.C.; Supervision, R.P.-V., L.T., A.R.-G., S.M., R.H. and J.C.; Validation, R.P.-V., L.T., A.R.-G., S.M., R.H. and J.C.; Visualization, R.P.-V., L.T., A.R.-G., S.M., R.H. and J.C.; Writing—original draft, R.P.-V., L.T., A.R.-G., S.M., R.H. and J.C.; Writing—review & editing, R.P.-V., L.T., A.R.-G., S.M., R.H. and J.C. All authors have read and agreed to the published version of the manuscript.

Funding: This work is a part of the eNMoLabs research project, and it is funded by the Universidad Nacional de Educación a Distancia (UNED).

Acknowledgments: The authors would like to acknowledge the support of the eNMoLabs research project for the period 2019–2020 from UNED; our teaching innovation group, CyberGID, started at UNED in 2018 and its associated CyberScratch PID project for 2020; another project for the period 2017–2018 from the Computer Science Engineering Faculty (ETSI Informática) in UNED; and the Region of Madrid for the support of E-Madrid-CM Network of Excellence (S2018/TCS-4307). The authors also acknowledge the support of SNOLA, officially recognized Thematic Network of Excellence (RED2018-102725-T) by the Spanish Ministry of Science, Innovation and Universities.

Conflicts of Interest: The authors declare no conflict of interest.

References

1. Al-Fuqaha, A.; Guizani, M.; Mohammadi, M.; Aledhari, M.; Ayyash, M. Internet of Things: A Survey on Enabling Technologies, Protocols, and Applications. *IEEE Commun. Surv. Tutor.* **2015**, *17*, 2347–2376. [CrossRef]

2. Kortuem, G.; Kawsar, F.; Sundramoorthy, V.; Fitton, D. Smart objects as building blocks for the Internet of things. *IEEE Internet Comput.* **2010**, *14*, 44–51. [CrossRef]

3. Shi, W.; Cao, J.; Zhang, Q.; Li, Y.; Xu, L. Edge Computing: Vision and Challenges. *IEEE Internet Things J.* **2016**, *3*, 637–646. [CrossRef]

4. Fog Computing and the Internet of Things: Extend the Cloud to Where the Things Are. Available online: https://www.cisco.com/c/dam/en_us/solutions/trends/iot/docs/computing-overview.pdf (accessed on 1 July 2020).

5. Chiang, M.; Zhang, T. Fog and IoT: An Overview of Research Opportunities. *IEEE Internet Things J.* **2016**, *3*, 854–864. [CrossRef]

6. Vaquero, L.M. EduCloud: PaaS versus IaaS Cloud Usage for an Advanced Computer Science Course. *IEEE Trans. Educ.* **2011**, *54*, 590–598. [CrossRef]

7. Xu, L.; Huang, D.; Tsai, W. Cloud-Based Virtual Laboratory for Network Security Education. *IEEE Trans. Educ.* **2014**, *57*, 145–150. [CrossRef]

8. AlHogail, A. Improving IoT Technology Adoption through Improving Consumer Trust. *Technologies* **2018**, *6*, 64. [CrossRef]

9. Pastor, R.; Romero, M.; Tobarra, L.; Cano, J.; Hernández, R. Teaching cloud computing using Web of Things devices. In Proceedings of the 2018 IEEE Global Engineering Education Conference, EDUCON 2018, Santa Cruz de Tenerife, Spain, 17–20 April 2018; pp. 1738–1745. [CrossRef]

10. Tobarra, L.; Robles-Gómez, A.; Pastor, R.; Hernández, R.; Cano, J.; López, D. Web of Things Platforms for Distance Learning Scenarios in Computer Science Disciplines: A Practical Approach. *Technologies* **2019**, *7*, 17. [CrossRef]

11. Shaikh, Y.; Parvati, V.K.; Biradar, S.R. Survey of Smart Healthcare Systems using Internet of Things (IoT). In Proceedings of the 2018 International Conference on Communication, Computing and Internet of Things (IC3IoT), Chennai, India, 15–17 February 2018; pp. 508–513. [CrossRef]

12. Qi, J.; Yang, P.; Xu, L.; Min, G. Advanced Internet of Things for Personalised Healthcare System: A Survey. *Pervasive Mob. Comput.* **2017**, *41*, 132–149. [CrossRef]

13. Yang, G.; Xie, L.; Mäntysalo, M.; Zhou, X.; Pang, Z.; Xu, L.; Kao-Walter, S.; Chen, Q.; Zheng, L.R. A Health-IoT Platform Based on the Integration of Intelligent Packaging, Unobtrusive Bio-Sensor and Intelligent Medicine Box. *IEEE Trans. Ind. Inf.* **2014**, *10*, 1. [CrossRef]

14. Mavrogiorgou, A.; Kiourtis, A.P.K.P.S.K.D. IoT in Healthcare: Achieving Interoperability of High-Quality Data Acquired by IoT Medical Devices. *Sensors* **2019**, *19*, 1978. [CrossRef] [PubMed]

15. Cruz-Piris, L.; Rivera, D.; Fernández, S.; Marsá-Maestre, I. Optimized Sensor Network and Multi-Agent Decision Support for Smart Traffic Light Management. *Sensors* **2018**, *18*, 435. [CrossRef] [PubMed]

16. Javed, M.A.; Ben Hamida, E.Z.W. Transport Systems for Smart Cities: From Theory to Practice. *Sensors* **2016**, *16*, 879. [CrossRef]

17. Liu, S.; Guo, L.; Webb, H.; Yao, X.; Chang, X. Internet of Things Monitoring System of Modern Eco-agriculture Based on Cloud Computing. *IEEE Access.* **2019**, *7*, 37050–37058. [CrossRef]

18. Elijah, O.; Abd Rahman, T.; Orikumhi, I.; Leow, C.Y.; Hindia, M. An Overview of Internet of Things (IoT) and Data Analytics in Agriculture: Benefits and Challenges. *IEEE Internet Things J.* **2018**, *5*, 3758–3773. [CrossRef]

19. Shaikh, Y.; Parvati, V.K.; Biradar, S.R. Connected vehicles and Internet of things. In Proceedings of the 2nd International Conference on Telecommunication and Networks (TEL-NET), Delhi, India, 10–11 August 2017; doi:10.1109/TEL-NET.2017.8343489. [CrossRef]

20. Singh, D.; Singh, M. Internet of vehicles for smart and safe driving. In Proceedings of the 2015 International Conference on Connected Vehicles and Expo (ICCVE), Shenzhen, China, 19–23 October 2015; pp. 328–329. [CrossRef]

21. Gartner Says 8.4 Billion Connected "Things" Will Be in Use in 2017, Up 31 Percent From 201 (Table 1). Available online: https://www.gartner.com/en/newsroom/press-releases/2017-02-07-gartner-says-8-billion-connected-things-will-be-in-use-in-2017-up-31-percent-from-2016 (accessed on 1 July 2020).

22. IoT Report: How Internet of Things Technology Growth Is Reaching Mainstream Companies and Consumers. Available online: https://www.businessinsider.com/internet-of-things-report?IR=T (accessed on 1 July 2020).

23. Clayer, J.; Toffolon, C.; Choquet, C. Patterns, Pedagogical Design Schemes and Process for Instructional Design. In Proceedings of the 2013 IEEE 13th International Conference on Advanced Learning Technologies, Beijing, China, 5–18 July 2013; pp. 304–306.

24. Li, C.; Xue, Y.; Wang, J.; Zhang, W.; Li, T. Edge-Oriented Computing Paradigms: A Survey on Architecture Design and System Management. *ACM Comput. Surv.* **2018**, *51*. [CrossRef]

25. Hunkeler, U.; Truong, H.L.; Stanford-Clark, A.J. MQTT-S - A publish/subscribe protocol for Wireless Sensor Networks. In Proceedings of the Third International Conference on COMmunication System softWAre and MiddlewaRE (COMSWARE 2008), Bangalore, India, 5–10 January 2008; pp. 791–798.

26. Armbrust, M.; Fox, A.; Griffith, R.; Joseph, A.D.; Katz, R.; Konwinski, A.; Lee, G.; Patterson, D.; Rabkin, A.; Stoica, I.; et al. A View of Cloud Computing. *Commun. ACM* **2010**, *53*, 50–58. [CrossRef]

27. Gusev, M.; Dustdar, S. Going Back to the Roots—The Evolution of Edge Computing, An IoT Perspective. *IEEE Internet Comput.* **2018**, *22*, 5–15. [CrossRef]

28. Toyama, S.; Hirayama, M. User Interface Design Method Considering UI Device in Internet of Things System. In Proceedings of the 2018 6th International Conference on Future Internet of Things and Cloud Workshops (FiCloudW), Barcelona, Spain, 6–8 August 2018; pp. 1–6. [CrossRef]

29. Ray, P.P. A Survey of IoT Cloud Platforms. *Fut. Comput. Informatics J.* **2016**, *1*, 35–46. [CrossRef]

30. Amazon AWS IoT Website. Available online: https://aws.amazon.com/es/iot/ (accessed on 1 July 2020).

31. Microsoft Azure IoT Website. Available online: https://azure.microsoft.com/es-es/overview/iot/ (accessed on 1 July 2020).

32. Watson IBM IoT Website. Available online: https://www.ibm.com/cloud/watson-iot-platform (accessed on 1 July 2020).

33. Postscapes Tech Website. Available online: https://www.postscapes.com/internet-of-things-platforms/ (accessed on 1 July 2020).

34. Fog Computing: A New Paradigm for IoT Clouds. Available online: https://www.ionos.es/digitalguide/servidores/know-how/fog-computing/ (accessed on 1 July 2020).

35. Shelby, Z.; Hartke, K.; Bormann, C. The Constrained Application Protocol (CoAP). *RFC* **2014**, *7252*, 1–112.

36. OASIS. Advanced Message Queuing Protocol (AMQP) Version 1.0; Part 3: Messaging. 2012. Available online: https://docs.oasis-open.org/amqp/core/v1.0/os/amqp-core-messaging-v1.0-os.html (accessed on 1 July 2020).

37. Bellavista, P.; Corradi, A.; Foschini, L.; Pernafini, A. Data Distribution Service (DDS): A performance comparison of OpenSplice and RTI implementations. *ISCC IEEE Comput. Soc.* **2013**, 377–383.

38. Saint-Andre, P. Extensible Messaging and Presence Protocol (XMPP): Core. *RFC* **2004**, *3920*.

39. Guerra, H.; Garcia, A.M.; Gomes, L.M.; Cardoso, A. An IoT remote lab for seismic monitoring in a programming course. In Proceedings of the 2017 4th Experiment@International Conference (exp.at'17), Faro, Algarve, Portugal, 6–8 June 2017; pp. 129–130.

40. Patil, S.; Supriya, K.; Uma, M.; Shettar, R.B.; Kumar, P. Open Ended Approach to Empirical Learning of IOT with Raspberry Pi in Modeling and Simulation Lab. In Proceedings of the 2016 IEEE 4th International Conference on MOOCs, Innovation and Technology in Education (MITE), Madurai, India, 7–10 December 2016; pp. 179–183.

41. Tunc, C.; Hariri, S.; De La Peña Montero, F.; Fargo, F.; Satam, P.; Al-Nashif, Y. Teaching and Training Cybersecurity as a Cloud Service. In Proceedings of the 2015 International Conference on Cloud and Autonomic Computing, Boston, MA, USA, 21–25 September 2015; pp. 302–308.

42. El-Hasan, T.S. Internet of Thing (IoT) Based Remote Labs in Engineering. In Proceedings of the 2019 6th International Conference on Control, Decision and Information Technologies (CoDIT), Paris, France, 23–26 April 2019; pp. 976–982.

43. Fernández-Pacheco, A.; Martin, S.; Castro, M. Implementation of an Arduino Remote Laboratory with Raspberry Pi. In Proceedings of the 2019 IEEE Global Engineering Education Conference (EDUCON), Dubai, UAE, 9–11 April 2019; pp. 1415–1418.

44. Leisenberg, M.; Stepponat, M. Internet of Things Remote Labs: Experiences with Data Analysis Experiments for Students Education. In Proceedings of the 2019 IEEE Global Engineering Education Conference (EDUCON), Dubai, UAE, 9–11 April 2019; pp. 22–27.

45. Rajurikar, N.S.; Kulkarni, S.V.; Patane, R.D. Implementation of centralized lab of an embedded web server using CoAP protocol on cloud computing. In Proceedings of the 2017 2nd IEEE International Conference on Recent Trends in Electronics, Information Communication Technology (RTEICT), Bengaluru, India, 19–20 May 2017; pp. 2267–2272.

46. Thibaud, M.; Chi, H.; Zhou, W.; Piramuthu, S. Internet of Things (IoT) in high-risk Environment, Health and Safety (EHS) industries: A comprehensive review. *Decis. Support Syst.* **2018**, *108*, 79–95. [CrossRef]

47. BitScope Website. Available online: https://www.bitscope.com/product/blade/ (accessed on 1 July 2020).

48. Docker Website. Available online: https://www.docker.com/ (accessed on 1 July 2020).

49. Merkel, D. Docker: Lightweight Linux Containers for Consistent Development and Deployment. *Linux J.* **2014**, *2014*.

50. Mora, H.; Signes Pont, M.T.; Gil, D.; Johnsson, M. Collaborative Working Architecture for IoT-Based Applications. *Sensors* **2018**, *18*, 1676. [CrossRef] [PubMed]

51. Coaty Website. Available online: https://coaty.io/ (accessed on 1 July 2020).

52. Docker Hub Website. Available online: https://hub.docker.com/ (accessed on 1 July 2020).

53. Docker Compose Website. Available online: https://docs.docker.com/compose/ (accessed on 1 July 2020).

54. Docker Swarm Website. Available online: https://docs.docker.com/engine/swarm/ (accessed on 1 July 2020).

55. Kubernetes Website. Available online: https://kubernetes.io/ (accessed on 1 July 2020).

56. Kubernetes Dashboard Website. Available online: https://kubernetes.io/docs/tasks/access-application-cluster/web-ui-dashboard/ (accessed on 1 July 2020).

57. IoT@UNED Lab Manager Portal. Available online: https://iot-at-uned.mybluemix.net/ (accessed on 1 July 2020).

58. Caminero, A.C.; Hernández, R.; Ros, S.; Tobarra, L.; Robles-Gómez, A.; San Cristóbal, E.; Tawfik, M.; Castro, M. Obtaining university practical competences in engineering by means of virtualization and cloud computing technologies. In Proceedings of the 2013 IEEE International Conference on Teaching, Assessment and Learning for Engineering (TALE), Kuta, Indonesia, 26–29 August 2013; pp. 301–306. [CrossRef]

59. Node-Red Website. Available online: https://nodered.org/ (accessed on 1 July 2020).

60. IBM IoT Watson Platform. Available online: https://cloud.ibm.com/docs/services/IoT?topic=iot-platform-getting-started (accessed on 1 July 2020).

61. IBM Cloudant website. Available online: https://www.ibm.com/cloud/cloudant (accessed on 1 July 2020).

62. Muthiah, A.; Ajitha, S.; Monisha Thangam, K.S.; Viveka Vikram, K.; Kavitha, K.; Ramalatha, M. Maternal ehealth Monitoring System using LoRa Technology. In Proceedings of the 2019 IEEE 10th International Conference on Awareness Science and Technology (iCAST), Marioka, Japan, 23–25 October 2019; pp. 1–4. [CrossRef]

63. Moustafa, H.; Schooler, E.M.; Shen, G.; Kamath, S. Remote monitoring and medical devices control in eHealth. In Proceedings of the 2016 IEEE 12th International Conference on Wireless and Mobile Computing, Networking and Communications (WiMob), New York, NY, USA, 17–19 October 2016; pp. 1–8. [CrossRef]

64. Liu, M.; Li, D.; Chen, Q.; Zhou, J.; Meng, K.; Zhang, S. Sensor Information Retrieval From Internet of Things: Representation and Indexing. *IEEE Access* **2018**, *6*, 36509–36521. [CrossRef]

65. Tobarra, L.; Robles-Gómez, A.; Pastor, R.; Hernández, R.; Duque, A.; Cano, J. Students' Acceptance and Tracking of a New Container-Based Virtual Laboratory. *Appl. Sci.* **2020**, *10*, 1091. [CrossRef]

66. Liu, I.F.; Chen, M.C.; Sun, Y.S.; Wible, D.; Kuo, C.H. Extending the TAM model to explore the factors that affect Intention to Use an Online Learning Community. *Comput. Educ.* **2010**, *54*, 600–610. [CrossRef]

67. Pastor, R.; Tobarra, L.; Robles-Gómez, A.; Cano, J.; Hammad, B.; Al-Zoubi, A.; Hernández, R.; Castro, M. Renewable energy remote online laboratories in Jordan universities: Tools for training students in Jordan. *Renew. Energy* **2020**, *149*, 749 – 759. [CrossRef]

Article

Block-Based Development of Mobile Learning Experiences for the Internet of Things

Iván Ruiz-Rube *, José Miguel Mota, Tatiana Person, José María Rodríguez Corral and Juan Manuel Dodero

School of Engineering, University of Cádiz, Avenida de la Universidad de Cádiz, 10, 11519 Puerto Real, Cádiz, Spain; josemiguel.mota@uca.es (J.M.M.); tatiana.person@uca.es (T.P.); josemaria.rodriguez@uca.es (J.M.R.C.); juanma.dodero@uca.es (J.M.D.)
* Correspondence: ivan.ruiz@uca.es

Received: 22 October 2019; Accepted: 9 December 2019; Published: 11 December 2019

Abstract: The Internet of Things enables experts of given domains to create smart user experiences for interacting with the environment. However, development of such experiences requires strong programming skills, which are challenging to develop for non-technical users. This paper presents several extensions to the block-based programming language used in App Inventor to make the creation of mobile apps for smart learning experiences less challenging. Such apps are used to process and graphically represent data streams from sensors by applying map-reduce operations. A workshop with students without previous experience with Internet of Things (IoT) and mobile app programming was conducted to evaluate the propositions. As a result, students were able to create small IoT apps that ingest, process and visually represent data in a simpler form as using App Inventor's standard features. Besides, an experimental study was carried out in a mobile app development course with academics of diverse disciplines. Results showed it was faster and easier for novice programmers to develop the proposed app using new stream processing blocks.

Keywords: Internet of Things (IoT); mobile apps; end-user development; App Inventor; block-based languages; map-reduce

---

1. Introduction

The Internet of Things (IoT) concept has several definitions, as involved technologies are continually evolving. IoT is defined as "a network that connects uniquely identifiable things to the Internet" [1]. These *things* have sensing and actuating capabilities and can be programmed, such that data can be collected and their state can change. IoT potentialities enable the development of a significant number of applications for improving citizens' life. Smart homes and buildings, smart cities, mobility and transportation, healthcare, agriculture and industry are some of the main areas of IoT application [1]. For a rapid materialization of IoT, the symbiosis among the physical world and the cyber world must be harmonious [2]. Interactions between humans and computing-enabled objects must be smarter and opportunistic [3]. As it may happen with humans' intelligence [4], the smartness of IoT things relies heavily on their sensory, interactive capabilities. In this vein, smart interactive objects enable creating tangible things to do different tasks in different application domains [5].

The development of smart IoT applications usually requires strong programming skills, which commonly exceed people's abilities. However, in recent years, several projects, such as Arduino and Raspberry Pi, aimed not only at professionals but also educators and students, have influenced the IoT expansion. These initiatives include both hardware platforms and programming tools, and a user community is growing around them.

Since the notation used in programming languages has a tremendous impact on novices [6], various tools to program IoT microcontrollers and microcomputers have emerged. These tools are based on block-based languages and proved to be useful for novice programmers. Learners of block-based languages depicted greater gains in algorithmic thinking [7] and a higher interest in computer science than those using text-based environments [8]. Differences between block-based languages and text-based languages often fade after learners transfer their acquired knowledge of computer programming to more professional, text-based languages and environments [9].

Currently, the most commonly used block-based programming tools, namely Scratch and App Inventor, provide capabilities to connect with external hardware devices, such as Arduino. However, they present some limitations when it comes to developing IoT applications, namely: (1) the absence of an easy mechanism for ingesting and processing event data streams and (2) the lack of usable mechanisms for visually representing data.

To facilitate authoring of IoT mobile apps, several visual components for a custom version of App Inventor, as well as a set of extensions for its block-based programming language, have been developed. With these components and language extensions, users can easily create apps that ingest data streams from available sensors, process them using a map-reduce programming style and then visualise the results of data processing graphically. The goal of this paper is to investigate how easy it is for non-experts to leverage such improved features to create their own smart IoT applications.

The block-based language approach followed in our research proposal has some limitations, which have been described in the literature. First, it may be applicable only for novice programmers who are learning to create their own smart IoT applications [9]. The research claims and results are not directly transferable to professional, text-based programming languages or even to other not block-based, visual programming paradigms [10]. Second, the use of programming concepts that are relevant to create smart IoT applications (such as state initialisation [11], parallelism [12], anonymous functions [13] and higher-order functions [14]) were adapted to visual and block-based languages. However, there are no evidences of learning improvements thanks to the use of such end-user development (EUD) approaches. Therefore, the use of block-based languages as an EUD approach for creating smart IoT applications may have limitations, which have to be overcome by more extensive research, as intended in this work.

The rest of the paper is structured as follows: the background and related works are presented in Section 2. Section 3 describes the main contribution. Two case studies are included in Sections 4 and 5. The former presents a usability study conducted with students of a computer programming fundamentals course whereas the latter was targeted at university lecturers. Finally, Section 6 discusses the results and draws the conclusions of this research.

2. Background & Related Works

IoT solutions are composed of hardware and software elements. Guth et al. [15] propose an IoT reference architecture from a comparison of various open-source (SiteWhere, OpenMTC and FIWARE) and proprietary (Amazon Web Services IoT) IoT platforms. Such architecture includes a set of sensors and actuators at the lower level. On the next level up, a hardware device is connected by a wired connection or wirelessly to sensors and actuators. Data communication protocols are required to manage the constraints of the smart devices, as well as gateways to translate data between different protocols and to forward communications. Middleware [15,16] processes the data received from the connected devices (e.g., by the execution of condition-action rules) to provide them to connected applications and sends commands to be executed by the corresponding actuators. Finally, IoT applications allow device-to-device and human-to-device interactions [17]. In the latter, mobile app-based smart interactive experiences can be provided for end-users.

Existing initiatives for learning and developing IoT solutions as well as block-based end-user development tools and their applications for creating IoT mobile experiences are described below.

2.1. Initiatives for Learning and Developing IoT Solutions

Arduino and Raspberry Pi are some of the most popular platforms used for educational purposes [1,18,19], with a huge development community. Arduino is a programmable circuit board, which can be connected with sensors and actuators of many types. Raspberry Pi is a single-board computer to run programs in a multitasking environment. However, the analog-digital conversion is not available onboard and thus additional hardware is required for interfacing with analog sensors such as photocells, joysticks and potentiometers.

Some initiatives and educational projects were carried out in order to teach IoT technologies for undergraduate and university students [20,21]. For example, in a project-based teaching and learning approach conceived for an IoT course [22], Raspberry Pi is used to devise and implement IoT designs. Other project-based learning courses for learning wired and wireless networking techniques have been offered to electrical and computer engineering students [23]. The use of microcontrollers with network connectivity and without complex operating systems provides cost-effective, well-supported and flexible platforms for developing IoT applications.

Moreover, the educational research outcome of teaching IoT device prototyping in a practical, real problem-based setting is presented [24] as a means for teaching computer science and software engineering. An example course outline for planning learning experiences in IoT prototyping is described along with a general assessment framework and best practice recommendations in order to facilitate personalised learning in analogous contexts.

Some educational approaches are based on the pocket labs (PL) concept to stimulate students' initiative and creativity. PL allow learners to experiment with real equipment in any place and at any time [25]. Despite that IoT and PL are not initially interrelated, the authors present a real case of IoT teaching practice based on Arduino and accompanying shields that includes sensors and actuators. PLs are combined with the online Tinkercad software tool to prototype and simulate electronic designs that include the Arduino boards.

Other initiatives for integrating IoT technologies in existing teaching-learning case studies were developed. For example, an IoT-based learning framework that integrates IoT and hardware/software technologies is used as part of a software engineering course for embedded system analysis and design [26]. Specifically, the authors introduced a lab development kit composed by Arduino and Raspberry Pi boards, sensors and XBee modules for providing wireless communication.

Common general-purpose programming languages can be used for developing IoT applications [19]. However, since IoT systems involve a wide variety of hardware and software components, depending on a range of distributed system and communication technologies, developing IoT applications is time-consuming and complex. Hence, a variety of IoT libraries, such as CoAPthon [27], and frameworks [28], such as IDeA, FRASAD, D-LITe, IoTLink, WebRTC based IoT application Framework, Datatweet, IoTSuite and RapIoT, have been developed to manage those complexities.

2.2. End-User Development Tools for IoT

Modern software programming tools hide much of the complexity of traditional programming languages. Recent *low code* software engineering approaches have been successful both for IoT [29] and for more general mobile application development [30]. Their general objective consists in making application creation easier for people without programming skills. This goal is shared by the research field known as end-user development (EUD). A recent review on this topic differentiates between end-user programming (EUP) and other software engineering activities that span the entire software development lifecycle [31]. The review was recently completed by another author, focusing on current EUD tools for developing IoT and robot applications [32].

Among EUP tools, block-based programming environment features are noteworthy [33] to enable composing programs without dealing with the syntactic issues of textual languages. Among such block-based languages and environments, Scratch [34,35] is very popular to create interactive games, stories and animations, as well as to share such creations on the Web. Scratch computer programs are built by dragging and dropping blocks that represent common programming elements, such as variables, expressions, conditions and statements. Another block-based EUP approach for robotic applications is Phratch, which is a Scratch-like live programming environment [36]. Besides, App Inventor [37,38] is an open-source block-based programming tool. This tool enables users without prior programming experience to create apps specifically for smartphones and mobile devices. In particular, it makes mobile app deployment easier for the end-user. Additionally to other tools' amenities, App Inventor allows end-users to perform interface design and software deployment tasks, which belong to the realm of EUD beyond EUP. End-users can drag, drop and arrange various interface and non-visible components through a visual designer and then use a block language editor to program the app logic behaviour in order to create and deploy fully functional mobile apps. App Inventor provides event handling as a form of trigger-action programming (TAP), which proved to be particularly suitable to define bespoke behaviours to respond to the multiple events that may occur in an IoT context [39]. End-users specify the behaviour of a system as events or triggers and response actions when the events occur [40].

Despite the availability of libraries and frameworks to work with IoT technologies, it is very complicated to find EUD solutions to assist non-IT professionals in a particular area or topic to develop their own IoT consumer applications and smart user experiences. For example, ScratchX [41] is an experimental platform that allows people to test experimental functionalities built by some developers for the Scratch visual language. These experimental extensions enable apps to integrate with web services and external hardware, such as Arduino or Raspberry Pi.

On the other hand, the MIT IoT App Inventor project [42] allows students, teachers and *makers* to implement IoT projects in the same way as they develop regular mobile apps. This project provides users with components and block extensions to read data from a great variety of sensors (e.g., moisture, pressure, temperature, noise, etc.) and control a multiplicity of actuators (e.g., buzzers, lights, motors, etc.) As apps run on mobile devices, they can take advantage of all built-in features provided by App Inventor, but they can also use the apps to interact with objects all around. Besides, UDOO [43] is a combined set of open hardware and software technologies to allow novice makers to create their own digital objects connected to the cloud and to define custom behaviour logic for sensors and actuators. In addition to the physical devices, UDOO includes an App Inventor extension to handle sensors and actuators from within mobile apps. Finally, IoT Inventor [44] is a web-based integration platform, not based on but inspired by App Inventor, with a friendly drag-and-drop composer interface to build personalised and reconfigurable services using smart IoT-enabled things.

All of the described extensions are targeted to handle sensors and actuators but they do not provide support for easily ingesting, processing and visualizing data.

3. Creating IoT Mobile Apps with VEDILS

VEDILS [45] is a visual environment for designing interactive learning scenarios. It is an authoring tool targeted at users without programming skills who want to create their own mobile apps. The platform is based on App Inventor, the programming tool to build apps for mobile devices. The current version requires Android devices, though an iOS-based version is currently being devised by MIT. The development environment relies on the Blockly library for its visual programming language based on blocks.

App Inventor provides several components for designing mobile apps' user interfaces as well as other features, including multimedia elements, communication with the device sensors, sharing through social networks, etc. In addition to the built-in components provided by App Inventor, VEDILS features new components to enrich the apps with virtual and augmented reality experiences and to serve multi-modal external Human Machine Interface (HMI) devices, such as hand gesture sensors or electroencephalography (EEG) headsets, among other features. The platform was also used to conduct a study on the suitability of visual languages for non-expert robot programmers [46].

Regarding IoT computing, several components and blocks were developed for VEDILS to ingest, process and visualise data from a diversity of sensors.

3.1. Ingesting IoT Data Streams

App Inventor manages the following block types for each component: property getters and setters (green blocks), functions (blue blocks) and event handlers (yellow blocks). VEDILS extends those with a particular type of block (similar to event handlers) for non-visual components that issue a continuous flow of data, as is the case of both internal and external sensors. These kinds of components (red blocks) provide the app developer with a data stream suitable to be treated with the processing blocks described in Section 3.2; these are triggered when data from the sensor are ingested for a predefined time window.

One of the most common ways of receiving data from an IoT sensor and sending commands to an actuator is via a Bluetooth connection. Thus, the built-in *BluetoothClient* component was extended with the new *StreamDataReceived* block (see Figure 1), which provides the data stream as well as a new *SecondsToGetStreamData* property to set the time period to fetch new data from the Bluetooth server.

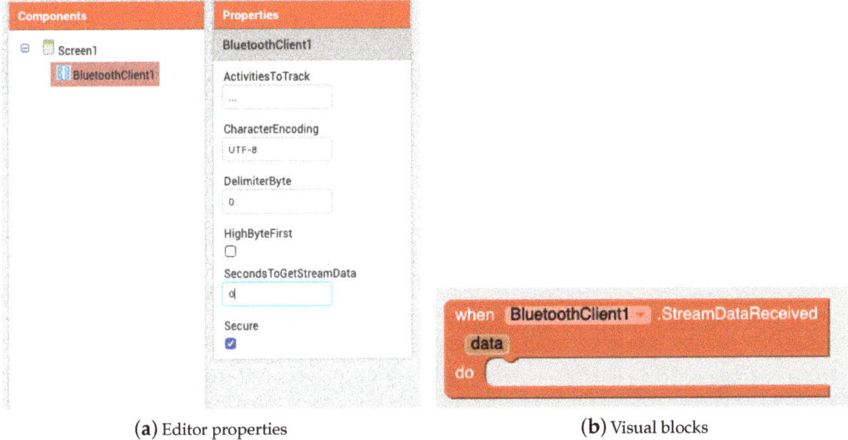

(a) Editor properties   (b) Visual blocks

Figure 1. Ingesting stream data from Bluetooth external devices.

In addition, every new VEDILS component that provides communication with internal or external devices can support the streaming blocks. For example, the *BrainwaveSensor* component, which enables to detect brain activity by means of an EEG headset, includes specific blocks for ingesting stream data from regular fast Fourier transform (FFT) bands (i.e., Theta, Alpha, Low Beta, High Beta and Gamma) of EEG channels (see Figure 2). It also includes a *TimeToStreamBandsData* property to set the time window. The current implementation works with the Emotiv Epoc+ and Insight headsets. Thus, this component enables a new range of mobile applications to monitor emotions, track cognitive performance and even control objects through learning a set of mental activity patterns that can be trained and interpreted as mental commands.

The blocks of Figure 2 were developed as Java class methods that support external and internal sensors. The data flow is internally managed as Java 8 streams. Besides, additional classes based on threads and the Java Timer API were required to periodically check data availability.

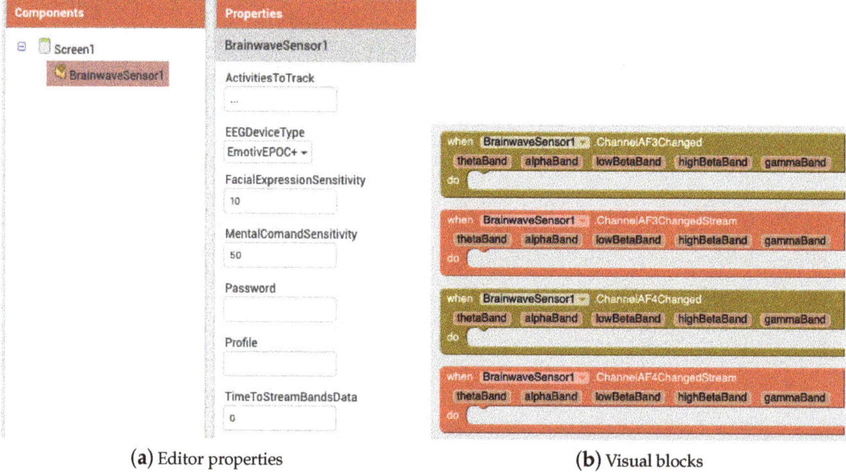

(a) Editor properties   (b) Visual blocks

Figure 2. Ingesting stream data from electroencephalography (EEG) headsets.

3.2. Processing IoT Data Streams

In order to address the issue of treating IoT data, several data processing blocks were developed and delivered with VEDILS (see Figure 3):

- *Filter* block: removes the elements that do not meet a specific condition from an input stream. For example, in a stream containing a set of numbers, developers could filter the odd numbers obtaining a new stream with only the even ones.
- *Map* block: applies an operation to each element of an input stream. For example, transforming a stream of lowercase words into a stream of uppercase words.
- *Reduce* (a) block: combines the elements contained in the input stream by applying the binary operator specified as a parameter. The combiner function must combine two numbers to return a new one, such as the maximum or minimum value.
- *Reduce* (b) block: combines the elements contained in the input stream, by applying one of the built-in mathematical operations. For example, computing the average or standard deviation of a 50-item stream.
- *Sort* (a) block: produces a new stream with the elements of the input stream according to the order induced by the comparator specified as a parameter. The comparator must be a function that returns a negative number if *item1* is less than *item2*; a positive one if *item1* is higher than *item2*; or zero if both items are equal.
- *Sort* (b) block: produces a new stream with the elements of the input stream according to its natural order, i.e., numerically or alphabetically. The block has a field to specify whether to apply an ascending or descending sorting.
- *Limit* block: shortens the stream size to the specified length. For example, collecting the first 10 items in the stream.

All the previous blocks are intermediary operations, except for the *Reduce* block, which is terminal. The intermediary blocks can be indistinctly chained, whereas the *Reduce* blocks must always appear on the left of the sequence of operations.

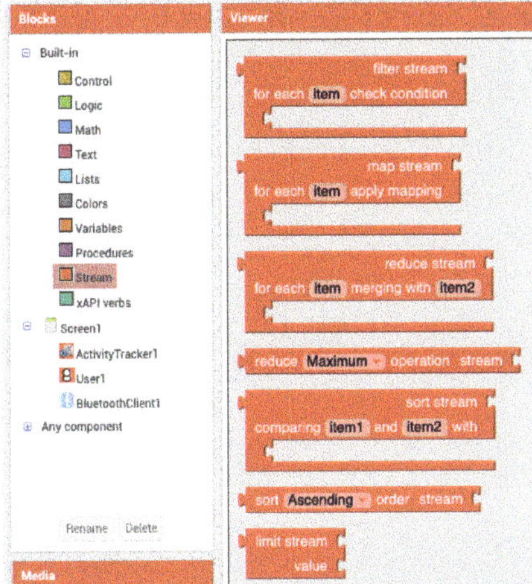

Figure 3. Visual blocks for processing stream data

Extensions for the visual programming language itself requires not only Java code but also other languages. The visual appearance of each block of the *Streams* palette is defined by a JavaScript fragment using the Blockly library API, whereas its run-time behaviour is defined by generating YAIL code. Young Android Intermediate Language (YAIL) is a set of abstractions for Kawa, a Java-based implementation of the Scheme functional language. Figures 4 and 5 show the code required to develop the *limit* block.

```
Blockly.Blocks['limit'] = {
    category: "Stream",
    init: function() {
    this.setColour(Blockly.STREAM_CATEGORY_HUE);
    this.appendValueInput('LIST')
        .setCheck(Blockly.Blocks.Utilities.YailTypeToBlocklyType("list",
        Blockly.Blocks.Utilities.INPUT))
        .appendField("limit stream")
        .setAlign(Blockly.ALIGN_RIGHT);
    this.appendValueInput("VALUE").setCheck(null).appendField('value').setAlign(Blockly.ALIGN_RIGHT);
    this.setTooltip('limit');
    this.setOutput(true);
    this.setPreviousStatement(false);
    },
    typeblock: [{ translatedName: 'limitBlock' }]
};
```

Figure 4. JS fragment for defining the *Limit* block with Blockly.

```
Blockly.Yail['limit'] = function() {
    // limit function
    var codeForList = Blockly.Yail.valueToCode(this, 'LIST', Blockly.Yail.ORDER_NONE) || null;
    var codeForBody = Blockly.Yail.valueToCode(this, 'VALUE', Blockly.Yail.ORDER_NONE) || null;
    var code = Blockly.Yail.YAIL_LIMIT + codeForBody +
        Blockly.Yail.YAIL_SPACER
    + codeForList + Blockly.Yail.YAIL_CLOSE_COMBINATION;
    return [ code, Blockly.Yail.ORDER_ATOMIC ];
};
```

Figure 5. JS fragment for generating the Young Android Intermediate Language (YAIL) code of the *Limit* block.

3.3. Visualising IoT Data Streams

App Inventor does not provide built-in capabilities to include charts or data tables in the apps. Thus, two new visible components were integrated into VEDILS for allowing developers to equip their apps with those kinds of visualisations (see Figure 6). The *Chart* component enables the creation of simple graphics such as bars, lines or pie charts, whereas the *DataTable* component is intended to present the data in a tabular format. Both components can be fed with a data stream from any sensor and they are customisable, for example, by configuring the category and value axes of charts.

(**a**) Editor properties   (**b**) Visual blocks

Figure 6. Visualizing stream data with the chart component

Both the *Chart* and *DataTable* components were developed as Java classes that inherit from the *AndroidViewComponent* superclass, already included in App Inventor. In run-time, these components provide an embedded *WebViewer* for the app screen, which points to an external HTML. That web page receives a JSON string containing the app data and renders it via the Google Chart API.

4. Evaluating IoT Mobile App Development with Students

This section presents a usability study of the VEDILS components for IoT computing. This test is aimed at checking whether these IoT components are suitable for learners when programming end-user IoT mobile apps. The test was defined and executed by following the guidelines provided by Rubin et al. [47].

4.1. Study Design

A three-hour workshop was conducted with students of a computer programming fundamentals course from a vocational education and training module. This workshop was implemented in the frame of a Code Week initiative (https://codeweek.eu/view/242496/desarrollo-sencillo-de-apps-moviles-para-iot). The research question for this study was the following: is it easier for students to develop IoT mobile apps using VEDILS than using App Inventor?. To evaluate the legibility degree and ease of use of the IoT blocks compared to App Inventor's, we designed the following experimental scenario: During the first hour, students learned about IoT, its basic concepts, components and architectures. Later on, they learned about the features and applications of the Raspberry Pi single-board computer and Sense HAT add-on. Then, they were presented with the running IoT sample app described below. The instructor taught students how to design the UI of the app as well as the blocks required to connect and disconnect the Raspberry Pi device through Bluetooth. Later on, the students were provided with a base project for both App Inventor and VEDILS to be completed during the rest of the session (see Figure 7a). Finally, they were asked to fill out a short questionnaire, including both quantitative and qualitative question items.

A quasi-experimental study was conducted with the students. One half of the participants created first the app with App Inventor and then with VEDILS; whereas the other half did it in reverse order.

To complete the app projects, the students were also provided with step-by-step tutorials to guide the development with both tools.

(a) (b)

Figure 7. Developing the mobile app for interacting with the external Internet of Things (IoT) device. (a) Students programming the mobile app; (b) Android app communicating with the Raspberry and its Sense HAT add-on.

4.2. The Sample App

The proposed app shows the average temperature received from the IoT sensor for the last 10 s. In order to bring the IoT app development closer to a more realistic scenario, the temperature measurements of the Raspberry Pi were generated from a simulation server. The python server running in the device generated random value series in the [14, 104] interval, measured as degrees Fahrenheit. Occasionally, abnormal values (i.e., 127°F) were generated to emulate measurement errors. The app had to obtain the temperature data, convert it into Celsius degrees, remove outliers and compute the average of the series.

4.2.1. User Interface Design

The app layout (see the picture in Figure 7b) is based on a vertical arrangement composed of several elements. At the top of the screen, there are buttons for connecting and disconnecting to/from the Raspberry Pi. A chart depicting the evolution of the temperature is included in the centre of the screen. At the bottom of the screen, there are buttons for sending commands to change the Sense HAT LED panel background colour. The LED panel in the picture shows the temperature measured every second.

From the user interface perspective, the only difference between the App Inventor and VEDILS versions is that the former requires a *Canvas* component, whereas the latter uses the new *Chart* component. Nevertheless, from the user programming perspective, there are some remarkable differences, as explained next.

4.2.2. Programming with App Inventor

A *Clock* component must be used to periodically check if there is new data to receive via the Bluetooth connection. Then, the developer must call the *ReceiveSignedBytes* method and then iterate through the data collection. For each individual value, a local variable is used to store the result of applying the conversion formula between the two measurement scales. Later on, a conditional statement must be applied to check if the calculated value is not an outlier. If so, that value must be added to an *accumulator* variable and increment by one the *counter* of valid measurements. Later on, the accumulator variable must be divided by the counter to compute the average (see Figure 8). Two text labels are accordingly updated to show the received raw of data and the computed average. Finally, the *updateChart* procedure is called to update the visual representation. Figure 9 shows how the *DrawLine* block in the *Canvas* must be used to depict the temperature along time as a line chart. This block requires a pair of (x,y) coordinates for both source and target points. Since the (0,0) point of the Canvas corresponds with its left upper corner, it is necessary to turn the temperature values into the proper values for the Y-axis. Besides, the X-axis must be consequently moved forward for each time instant. Some other variables must also be used to control the coordinates. Furthermore, at the beginning of the drawing process and every time the canvas right-edge is reached, the drawing area must be cleared and some horizontal lines must be drawn to represent certain temperature milestones (0°, 10°, etc.). Besides, some variables must be reinitialised.

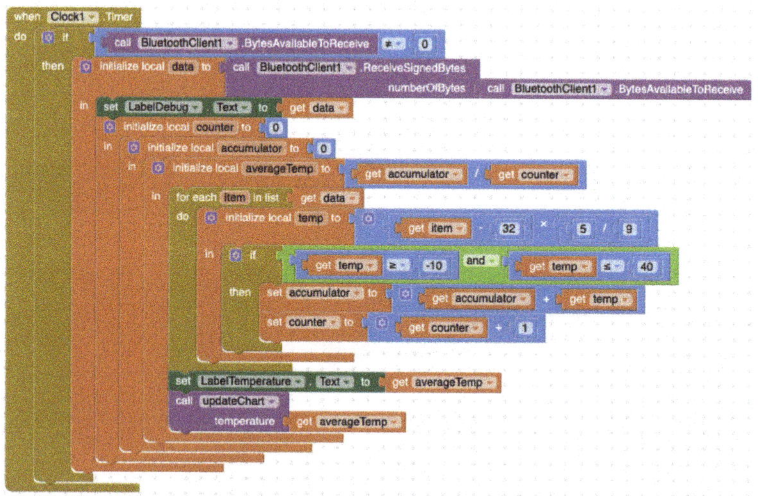

Figure 8. Processing temperature data with App Inventor.

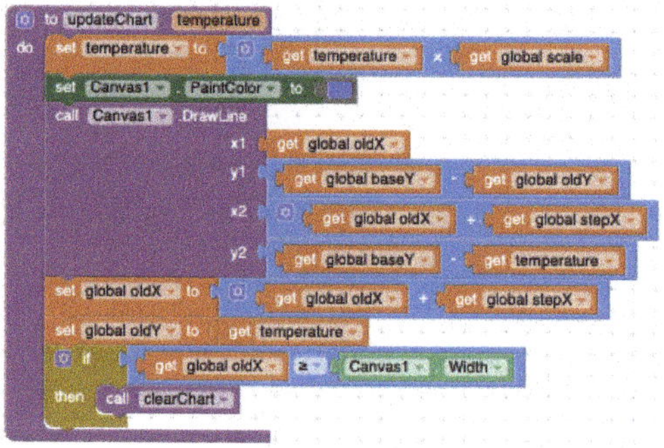

Figure 9. Drawing temperature data with App Inventor.

4.2.3. Programming with VEDILS

With VEDILS (see Figure 10), the developer must handle the *StreamDataReceived* block, which directly provides a data stream of temperature measurements. This data stream is pipelined through a series of processing steps. With the *Map* block, every item in the stream is mapped into its corresponding Fahrenheit value; with the *Filter* block, outliers are discarded according to the validity condition; and finally, with the *Reduce* block, the data stream is summarized by computing an average. The *AppendData* block is used to depict the average temperature along time in the line chart. Thus, the computed value must be sent to the *Chart* component, together with the current timestamp provided by *Clock*. Previously, the *Chart* must have been configured with the category and value axes (see Figure 11).

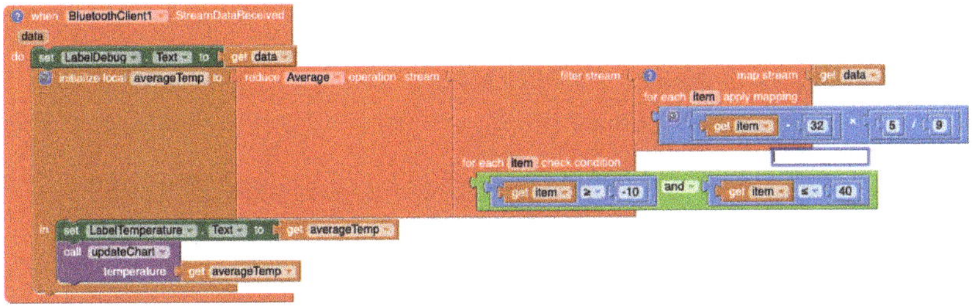

Figure 10. Processing temperature data with VEDILS.

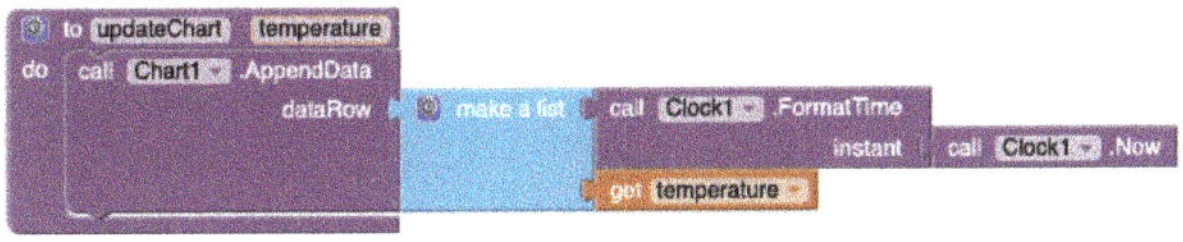

Figure 11. Drawing temperature data with VEDILS.

4.3. Data Compilation

The data collection was performed without interacting with the subjects during the experiment (i.e., an indirect method). The online questionnaire designed for the survey includes, in addition to the consent form, several questions to determine the initial status of participants as well as to compile the students' opinions after the test. They were asked about their expertise level creating software programs with visual languages and with text-based programming languages. Regarding the post-test, some questions related to the perceived ease of use of App Inventor and VEDILS were included. They deal with the tool usability for (i) connecting/disconnecting via Bluetooth and sending commands to the IoT sensor; (ii) consuming temperature data from the sensor, applying a transformation for changing their measurement scale, removing the outliers and computing the average; and (iii) drawing a chart with the temperature evolution. The answers to these questions follow a five-level Likert scale (1-Strongly disagree, 2-Disagree, 3-Neither agree nor disagree, 4-Agree and 5-Strongly agree). The participants were also asked about their intention to use App Inventor or VEDILS to create more IoT projects. Study data as well as the resources used are linked in the Supplementary Materials.

4.4. Analysis and Findings

This study aimed at checking whether it is easier for students to develop the proposed IoT mobile app using VEDILS rather than with App Inventor. Ten students (eight men and two women) aged 24 (stddev = 2) participated in the study. All of them were first-year students of a vocational course in web development. By the time the workshop was conducted, they had not yet learned any textual programming tool. They had only studied and used the App Inventor platform. Only one of them had previous experience with traditional text-based coding languages.

All students agreed that it is easy (avg = 4.0, stddev = 0.0) to develop the routines for connecting/disconnecting via Bluetooth and sending commands to the IoT sensor with App Inventor and VEDILS (both tools share the same blocks for that purpose). Regarding the temperature data ingestion and processing steps, most students neither agree nor disagree (avg = 2.77, stddev = 0.40) that these steps are easy to develop with App Inventor. Nevertheless, most students agreed that it is easy to develop these routines with VEDILS (avg = 3.88, stddev = 0.26). This perceived ease of use is even more substantial when developing the temperature evolution chart: the App Inventor *Canvas* component (avg = 2.66, stddev = 0.44) versus the VEDILS *Chart* component (avg = 4, stddev = 0.23). Finally, regarding the question about which tool they would use to develop mobile apps that consume data from IoT sensors and depict them in a chart, 66.6% of the participants chose VEDILS, 11.1% chose App Inventor, whereas the rest did not indicate a preference. In short, all participants rated VEDILS better than App Inventor, except for the student who already had coding skills.

A total of 177 blocks were required for developing the App Inventor version of the app, whereas only 84 were required in VEDILS. In both cases, procedures were used to avoid as much as possible the number of duplicate blocks. Accordingly, the difference in the size of the projects may relate to the students' perceived ease of use for both tools. That difference is particularly pronounced when it comes to presenting the temperature chart because developers must handle many details of the drawing process. The results are also consistent with the qualitative opinions expressed by the students, who highlighted the saving of programming effort required to consume, process and visualise IoT data thanks to the abstractions provided by VEDILS.

5. Evaluating VEDILS Data Processing Blocks with Academics

While the above section evaluates the components and language extensions provided by VEDILS for IoT computing, this case study solely focuses on the data processing blocks. The main objective is to check the development agility and the usability of the stream blocks compared to the standard built-in blocks for processing data. The design, implementation and analysis of the experimental study are presented below.

5.1. Study Design

The study was performed through six editions of an introductory course of mobile app development with App Inventor/VEDILS between January and February 2018. These courses are part of the Cádiz university's docent innovation program, in which several IT-related courses are regularly delivered to their associated lecturers and researchers.

The reference framework for establishing the hypotheses of this study is based on the potential benefits of certain computer programming paradigms over others [48]. Some authors explored techniques for introducing parallelism concepts, anonymous procedures and higher-order functions into block languages [12–14]. In this particular case of application development, we analyse the ease and agility of using block-based versions of the map-reduce constructs from the functional programming paradigm versus the iterative constructs (i.e., loops) from the imperative programming paradigm. The research questions posed for this study are the following: RQ1—Is there any difference in users' perception of the complexity of the stream processing blocks? RQ2—Is it easier for users to develop apps that collect and process data samples using functional blocks rather than using imperative blocks? and RQ3—Is it faster for users to develop apps that collect and process data samples using functional blocks rather than using imperative blocks?

To find answers to the research questions, the following scenario was carried out. First, all the academics interested in enrolling in the course were arbitrarily allocated in one of the (six) course editions. Each course lasted five hours and the participants were first taught with a short introduction to the educational applications of mobile devices. Next, the instructors explained the fundamentals of visual programming and the VEDILS tool's features.

Second, to reinforce and consolidate what was learned, participants created a number of educational mobile apps. These apps leverage the smartphone sensory and multimedia elements provided by App Inventor as well as the augmented reality capabilities provided by VEDILS. During the course, all the participants had to develop the same apps, except for one that emulates dice rolling. In addition to simulating the dice, in three of the course editions attendants who represented the control groups had to include an additional routine to calculate the count of odd numbers in a sequence of dice roll samples, whereas in the other three editions, attendants who represented the experimental groups had to program the count of even numbers. For the control groups, the attendants were accordingly taught about the loop statements for data processing, whereas for the experimental groups, the participants were taught about stream blocks.

Finally, the course attendants were asked to develop a citizen science mobile app by themselves. In this vein, smartphones enable to automate data collection and enrich observations with photographs, sound recordings and global positioning system (GPS) coordinates using embedded sensors [49]. The app requirements were: (i) to simulate the input of a numerical measurement of an external phenomenon and (ii) to compute the average of the collected measurements, excluding values out of a permitted value range.

The development of the citizen science app was required to obtain the course completion certificate. The assignment delivery was due within two weeks of course completion. In addition to submitting the developed apps, an online questionnaire had to be filled out. Answers to the questionnaire were analysed using quantitative techniques.

5.2. Data Compilation

A total of 45 users attended the VEDILS course. Data collection was performed without interacting with the subjects through an online questionnaire. The survey included questions related to the participants' knowledge area, age, gender, years of teaching and research experience, highest academic degree obtained and prior expertise in creating computer programs with a visual and/or text-based programming language. Regarding the post-test, questions related to the perceived ease of use of App Inventor and VEDILS were included. These questions pointed to several aspects, such as the use of variables and data lists, control flow statements and loop blocks (for the control groups) and the use of stream blocks (for the experimental groups). In addition, similar questions were included to check the participants' self-confidence when developing the citizen science app. The answers to all the questions were on a five-level Likert scale.

Besides, all the app project files submitted to the learning management store (i.e., Moodle) for the instructor's review were subsequently processed through a data integration process for analytic purposes. Among other data, the following were automatically extracted: time spent to develop the app, the number of blocks used, number of debugs and compilations required to complete the app. Study data as well as the resources used are linked in the Supplementary Materials.

5.3. Analysis and Findings

The 45 participants (17 women and 28 men) were, on average, 41 years old, had 13 years of teaching experience and 11 years of researching experience. Furthermore, 62.22% of the academics had a Ph.D. Their background is as follows: Arts and Humanities (2.22%), Computer Science (20%), Engineering and Architecture (4.44%), Health Sciences (17.78%), Laws and Social Sciences (28.89%) and Natural Sciences (26.67%). In terms of their previous programming experience, from nothing (1) to expert (5), they had scarce visual (avg = 1.82) and textual (2.15) programming skills. Overall, 25 subjects were part of the control groups, whereas the experimental groups were composed of 20 subjects.

Tables 1 and 2 show the users' perception of the stream processing blocks complexity and the ease of development of the app created to collect and process data samples. Data are grouped in the table according to the participants' gender, academic degree, knowledge area and previous experience with visual and textual programming languages.

Table 1. Results of the survey with academics: perceived ease of use (the italic font shows the average and chi-squared values whereas the bold one indicates significant differences).

User Profile	Loop Blocks	Stream Blocks	Average	Chi-Squared
Gender				
Man	3.71	3.69	*3.71*	*0.72*
Woman	3.73	4.40	*3.94*	*0.55*
Chi-squared	*0.61*	*0.41*		
Academic degree				
Non-doctorate	3.77	4.00	*3.88*	*0.09*
Doctorate	3.68	3.82	*3.75*	*0.24*
Chi-squared	*0.46*	***0.04***		
Knowledge area				
EHSE	4.00	3.64	*4.00*	*0.48*
SSH	3.22	4.75	*3.22*	*0.29*
Chi-squared	*0.46*	*0.17*		
Experience with visual programming languages				
Non-experienced	3.89	4.00	*3.67*	*0.31*
Experienced	4.57	3.60	*4.17*	***0.02***
Chi-squared	*0.19*	***0.01***		
Experience with textual programming languages				
Non-experienced	3.21	3.83	*3.54*	*0.58*
Experienced	4.36	4.00	*4.24*	*0.46*
Chi-squared	*0.17*	*0.32*		
All academics				
Academics	*3.72*	*3.89*	*3.80*	*0.83*

Table 2. Results of the survey with academics: ease of development of the app (the italic font shows the average and chi-squared values whereas the bold one indicates significant differences).

User Profile	Loop-Based	Stream-Based	Average	Chi-Squared
Gender				
Man	3.07	4.08	*3.57*	*0.17*
Woman	2.81	3.33	*3.00*	*0.37*
Chi-squared	*0.48*	***0.05***		
Academic degree				
Non-doctorate	2.88	4.25	*3.53*	*0.31*
Doctorate	3.00	3.54	*3.25*	*0.60*
Chi-squared	*0.23*	*0.32*		
Knowledge area				
EHSE	3.37	3.93	*3.65*	*0.63*
SSH	2.22	3.60	*2.71*	*0.18*
Chi-squared	*0.23*	*0.29*		
Experience with visual programming languages				
Non-experienced	2.55	3.71	*3.09*	***0.05***
Experienced	4.00	4.20	*4.08*	*0.48*
Chi-squared	*0.06*	*0.52*		
Experience with textual programming languages				
Non-experienced	2.07	3.54	*2.82*	***0.01***
Experienced	4.09	4.50	*4.24*	*0.54*
Chi-squared	***0.00***	*0.11*		
All academics				
Academics	*2.96*	*3.84*	*3.36*	*0.13*

Concerning the participants' gender, women perceived that the stream blocks were easier to use (avg = 4.4) than for men (avg = 3.69) but interestingly enough, men were the ones who found the app development more comfortable with those blocks (man's avg = 4.08 vs. woman's avg = 3.33). Furthermore, non-doctorates found the development much easier with stream blocks (avg = 4.25) than the traditional ones (avg = 2.88). Besides, Social Sciences and Humanities (SSH) academics perceived the stream blocks easier to use (avg = 4.75) compared to the Earth & Health Sciences and Engineering (EHSE) lecturers (avg = 3.64). SSH academics also found it difficult (avg = 2.22) to develop the app with the loop blocks, whereas they did not have that much trouble with the stream ones (avg = 3.6).

It is interesting to note ($p < 0.05$) that users with previous experience in visual programming languages perceived loop blocks (avg = 4.57) easier than stream blocks (avg = 3.56). Nevertheless, there is a significant difference ($p < 0.05$) in the fact that academics without experience with visual languages developed the app easier with the map-reduce blocks (avg = 3.71) than with the standard loop blocks (avg = 2.55). As expected, there is also a significant difference ($p < 0.05$) concerning the ease of development of the proposed app with the map-reduce blocks (avg = 3.54) compared to the standard loop blocks (avg = 2.07) for academics without experience with textual languages.

Regarding the apps the lecturers had to create as final assignment of the course, 39 out 45 were correctly developed: 16 apps use the traditional loop blocks and 23 use the stream blocks. Table 3 shows the direct metrics obtained from the app projects. As can be observed, all the apps which had to be developed with stream blocks were completed. The remaining six apps were expected to be developed using the traditional loop blocks. On average, three hours were needed to develop the app with the standard loop blocks, whereas fewer than two hours were required to create the same app with the new stream blocks. That is also tested with a significant difference ($p < 0.05$). Furthermore, the average number of builds and debugs performed for the stream-based apps is fewer than for the loop-based one.

Table 3. Indicators of the developed apps (the italic font shows the average and Mann-Whitney U Test values whereas the bold one indicates significant differences).

		% Completion	Minutes Spent	Number of Debugs + Builds
Loop-based		72%	180.71	13.25
Stream-based		100%	111.97	9.74
	Average	*86.66%*	*140.17*	*11.18*
	Mann-Whitney U Test		***0.024***	*0.16*

To sum up, with regard to the question (RQ1), there is no difference in the users' perception of the complexity of the stream processing blocks (avg=3.89) and the loop blocks (avg = 3.72). Concerning whether it is easier for users to develop apps which collect and process data samples with functional blocks rather than with imperative blocks (RQ2), the participants agreed that the development of the requested app was easier with the map-reduce blocks (avg = 3.84) than with loop ones (avg = 2.96). Finally, the indicators obtained for RQ3 point that it is faster for users (100% completion of the projects, a fewer number of debugs required to develop the app and a significant difference (around 38%) in saving development time) to collect and process data samples with functional blocks rather than with imperative blocks.

6. Discussion and Conclusions

Developing smart user experiences based on IoT technologies is a very complicated task, especially for non-IT professionals. To address these barriers, some of the popular block-based tools aimed at learners in computer programming (e.g., Scratch or App Inventor) were extended with modules to communicate with external hardware. However, they do not provide adequate support for easily ingesting, processing and visualising data from sensors.

In this research, some components and blocks developed explicitly for a custom version of App Inventor, called VEDILS, were proposed. They are devised to facilitate the ingestion of data from sensors in time intervals, to process received data by using a pipelined sequence of mapping, filtering and reducing operations, and finally, to represent them graphically or in a tabular format.

Two studies, namely a quasi-experimental study conducted with students and an experimental with academics, were conducted to evaluate the contribution. From the first study, students considered that it was easier for them to develop the routines for ingesting, processing and visualising data from the external temperature sensor with the VEDILS IoT features rather than with the equivalent components and blocks in App Inventor. From the second study, aimed at only checking the data processing blocks, participants did not perceive stream blocks easier than the loop blocks. Nevertheless, with statistical significance, it was faster for academics to develop the proposed app with the stream blocks, and easier specifically for novice programmers.

Additionally, threats to validity must be taken into account. To maximise the internal validity and the construct validity, we maintained a detailed protocol for both studies. Peer researchers reviewed them, and actions were considered to minimise bias. In the first study, the students completed the app development projects for both App Inventor and VEDILS but in reverse order to minimise the learning effect on the subjects. In the study conducted with academics, they were randomly distributed into different groups. In addition, every course edition was taught with the same instructors (also the authors of this paper). Furthermore, all course attendants were required to develop the same app with the same requirements to ensure the count of minutes spent, debugs and builds required to create the apps were not affected by other factors. Apart from the data automatically extracted from the developed projects (time spent, number of builds and number of debugs), the rest of the variables used for our experiments to measure user perceptions are subjective so that they can also be considered as validity threats.

The limited size of the student sample can be viewed as an external validity threat. Moreover, although the second study has a user sample more extensive than the first one, it is only aimed at academics. As a result, we cannot assure that the obtained findings can be generalised to professionals of other disciplines or conventional users. Hence, more experimentation and analysis are required to evaluate to what extent the findings presented in this work are of relevance for other cases.

It is necessary to consider the limitations of the current work. First, since the new type of language block for providing app developers with a data stream according to a predefined period of time was only incorporated for the standard *BluetoothClient* component and the *BrainwaveSensor* of VEDILS, it is not currently possible to harness it for other built-in App Inventor sensors. In addition, the extension component for using Bluetooth Low Energy (BLE) technology, which is not part of the App Inventor main distribution, is not yet supported for our contribution. Second, the current implementation of the components for visualising data do not allow developers to customise colours, lines widths, font sizes, etc., which are format aspects usually required when designing charts.

Smart homes and buildings, smart cities, mobility and transportation, healthcare, agriculture and industry are some of the main areas of IoT application [1]. The study conducted with students illustrated the potential application of our approach to smart buildings, e.g., for monitoring room temperature. Nevertheless, the contribution presented in the paper is expected to be useful to create IoT applications for other areas. Thus, for example, non-expert programmers (researchers, patients and healthcare professionals) will be able to develop apps for wellness and healthcare purposes without struggling with the complexities of the common mobile programming languages, namely Java or Swift. These kinds of apps are usually data-intensive and require to process users' biometric data, which is ingested from wearable devices, such as smart bands or chest straps, among others.

This research tried to investigate whether the components and extensions presented in the paper contribute to the popularisation of IoT-based mobile app development. In this vein, EUD platforms and, in particular, enriched block-based authoring tools as VEDILS, can simplify development tasks of novice end-user programmers. Furthermore, according to the obtained results, the use of blocks based on the map-reduce paradigm from functional programming streamlines the development of data processing functions in IoT consumer apps, although more experimentation is required. As future work, we plan to support the BLE extension for App Inventor to improve the customising features of the *Chart* and *DataTable* components.

Supplementary Materials: The software for authoring IoT apps with data processing and visualising capabilities can be used through the VEDILS web site http://vedils.uca.es/. Besides, to guarantee the reproducibility of the studies, all the resources developed are available online at http://www.mdpi.com/1424-8220/19/24/5467/s1. This link provides a ZIP package containing: questionnaires and results of both studies, PDF presentations (in Spanish) for the course with academics and the workshop with students and solutions for the different exercises and tutorials.

Author Contributions: Conceptualization, I.R.-R.; methodology, I.R.-R. and J.M.D.; software, T.P. and I.R.-R.; validation, J.M.M.; formal analysis, I.R.-R. and J.M.M.; investigation, I.R.-R., J.M.M., and J.M.R.C.; resources, I.R.-R. and J.M.D.; data curation, I.R.-R. and J.M.M.; writing—original draft preparation, I.R.-R., J.M.M. and J.M.R.C.; writing—review and editing, I.R.-R. and J.M.D.; visualization, I.R. and J.M.M.; supervision, I.R.-R.; project administration, J.M.D.; funding acquisition, J.M.D.

Funding: This work was developed in the VISAIGLE project, funded by the Spanish National Research Agency (AEI) with ERDF funds under grant ref. TIN2017-85797-R.

Conflicts of Interest: The authors declare no conflict of interest. The funders had no role in the design of the study; in the collection, analyses, or interpretation of data; in the writing of the manuscript, or in the decision to publish the results.

Abbreviations

The following abbreviations are used in this manuscript:

BLE	Bluetooth Low Energy
EEG	electroencephalography
EHSE	Earth & Health Sciences and Engineering
EUD	end-user development
EUP	end-user programming
FFT	fast Fourier transform
GPS	global positioning system
HMI	Human Machine Interface
IoT	Internet of Things
MIT	Massachusetts Institute of Technology
PL	pocket lab
SSH	Social Sciences and Humanities
TAP	trigger-action programming
VEDILS	Visual Environment for Designing Interactive Learning Scenarios
YAIL	Young Android Intermediate Language

References

1. Hassan, Q.F.; Madani, S.A. *Internet of Things: Challenges, Advances, and Applications*; CRC Press: Boca Raton, FL, USA, Taylor & Francis Group, LLC: Abingdon, UK, 2017.
2. Zhong, N.; Ma, J.; Huang, R.; Liu, J.; Yao, Y.; Zhang, Y.; Chen, J. Research Challenges and Perspectives on Wisdom Web of Things (W2T). In *Wisdom Web of Things*; Springer International Publishing: Cham, Switzerland, 2016; pp. 3–26. [CrossRef]
3. Guo, B.; Zhang, D.; Wang, Z.; Yu, Z.; Zhou, X. Opportunistic IoT: Exploring the harmonious interaction between human and the internet of things. *J. Netw. Comput. Appl.* **2013**, *36*, 1531–1539. [CrossRef]
4. Näätänen, R.; Tervaniemi, M.; Sussman, E.; Paavilainen, P.; Winkler, I. 'Primitive intelligence' in the auditory cortex. *Trends Neurosci.* **2001**, *24*, 283–288. [CrossRef]
5. Ardito, C.; Desolda, G.; Lanzilotti, R.; Malizia, A.; Matera, M. Analysing trade-offs in frameworks for the design of smart environments. *Behav. Inf. Technol.* **2019**, 1–25. [CrossRef]
6. Stefik, A.; Siebert, S. An Empirical Investigation into Programming Language Syntax. *ACM Trans. Comput. Educ.* **2013**, *13*. [CrossRef]

7. Grover, S.; Pea, R.; Cooper, S. Designing for deeper learning in a blended computer science course for middle school students. *Comput. Sci. Educ.* **2015**, *25*, 199–237. [CrossRef]

8. Weintrop, D.; Wilensky, U. Comparing Block-Based and Text-Based Programming in High School Computer Science Classrooms. *ACM Trans. Comput. Educ.* **2017**, *18*. [CrossRef]

9. Weintrop, D.; Wilensky, U. Transitioning from introductory block-based and text-based environments to professional programming languages in high school computer science classrooms. *Comput. Educ.* **2019**, *142*. [CrossRef]

10. Paternò, F. End user development: Survey of an emerging field for empowering people. *ISRN Softw. Eng.* **2013**, *2013*, 532659. [CrossRef]

11. Franklin, D.; Hill, C.; Dwyer, H.; Hansen, A.; Iveland, A.; Harlow, D. Initialization in Scratch: Seeking Knowledge Transfer. In Proceedings of the 47th ACM Technical Symposium on Computing Science Education, Memphis, TN, USA, 2–5 March 2016; pp. 217–222. [CrossRef]

12. Bogaerts, S. Hands-On Exploration of Parallelism for Absolute Beginners with Scratch. In Proceedings of the 2013 IEEE International Symposium on Parallel Distributed Processing, Workshops and Phd Forum, Cambridge, MA, USA, 20–24 May 2013; pp. 1263–1268. [CrossRef]

13. Harvey, B.; Mönig, J. Lambda in blocks languages: Lessons learned. In Proceedings of the 2015 IEEE Blocks and Beyond Workshop (Blocks and Beyond), Atlanta, GA, USA, 22 October 2015; pp. 35–38.

14. Kim, S.; Turbak, F. Adapting higher-order list operators for blocks programming. In Proceedings of the 2015 IEEE Symposium on Visual Languages and Human-Centric Computing (VL/HCC), Atlanta, GA, USA, 18–22 October 2015; pp. 213–217.

15. Guth, J.; Breitenbücher, U.; Falkenthal, M.; Leymann, F.; Reinfurt, L. Comparison of IoT platform architectures: A field study based on a reference architecture. In Proceedings of the 2016 Cloudification of the Internet of Things (CIoT), 2016, Paris, France, 23–25 November 2016; pp. 1–6.

16. Ngu, A.H.; Gutierrez, M.; Metsis, V.; Nepal, S.; Sheng, Q.Z. IoT middleware: A survey on issues and enabling technologies. *IEEE Internet Things J.* **2016**, *4*, 1–20. [CrossRef]

17. Lee, I.; Lee, K. The Internet of Things (IoT): Applications, investments, and challenges for enterprises. *Bus. Horizons* **2015**, *58*, 431–440. [CrossRef]

18. McEwen, A.; Cassimally, H. *Designing the Internet of Things*; John Wiley & Sons: Indianapolis, IN, USA, 2013.

19. Singh, K.J.; Kapoor, D.S. Create Your Own Internet of Things: A survey of IoT platforms. *IEEE Consum. Electron. Mag.* **2017**, *6*, 57–68. [CrossRef]

20. Ali, F. Teaching the internet of things concepts. In Proceedings of the WESE'15: Workshop on Embedded and Cyber-Physical Systems Education, Amsterdam, The Netherlands, 4–9 October 2015; p. 10.

21. Raikar, M.M.; Desai, P.; Vijayalakshmi, M.; Narayankar, P. Upsurge of IoT (Internet of Things) in engineering education: A case study. In Proceedings of the 2018 International Conference on Advances in Computing, Communications and Informatics (ICACCI), Bangalore, India, 19–22 September 2018; pp. 191–197.

22. Zhong, X.; Liang, Y. Raspberry Pi: An effective vehicle in teaching the internet of things in computer science and engineering. *Electronics* **2016**, *5*, 56. [CrossRef]

23. He, N.; Bukralia, R.; Huang, H.W. Teaching wireless networking technologies in the internet-of-things using ARM based microcontrollers. In Proceedings of the 2017 IEEE Frontiers in Education Conference (FIE), Indianapolis, IN, USA, 18–21 October 2017; pp. 1–4.

24. Mäenpää, H.; Varjonen, S.; Hellas, A.; Tarkoma, S.; Männistö, T. Assessing IoT projects in university education: A framework for problem-based learning. In Proceedings of the 39th International Conference on Software Engineering: Software Engineering and Education Track, Buenos Aires, Argentina, 20–28 May 2017; pp. 37–46.

25. Cvjetkovic, V. Pocket labs supported IoT teaching. *Int. J. Eng. Pedagog.* **2018**, *8*, 32–48. [CrossRef]

26. He, J.; Lo, D.C.T.; Xie, Y.; Lartigue, J. Integrating Internet of Things (IoT) into STEM undergraduate education: Case study of a modern technology infused courseware for embedded system course. In Proceedings of the 2016 IEEE Frontiers in Education Conference (FIE), Erie, PA, USA, 12–15 October 2016; pp. 1–9.

27. Tanganelli, G.; Vallati, C.; Mingozzi, E. CoAPthon: Easy development of CoAP-based IoT applications with Python. In Proceedings of the 2015 IEEE 2nd World Forum on Internet of Things (WF-IoT), Milan, Italy, 14–16 December 2015; pp. 63–68.

28. Udoh, I.S.; Kotonya, G. Developing IoT applications: Challenges and frameworks. *IET Cyber-Phys. Syst. Theory Appl.* **2018**, *3*, 65–72. [CrossRef]

29. Pantelimon, S.G.; Rogojanu, T.; Braileanu, A.; Stanciu, V.D.; Dobre, C. Towards a Seamless Integration of IoT Devices with IoT Platforms Using a Low-Code Approach. In Proceedings of the IEEE 5th World Forum on Internet of Things, Limerick, Ireland, 15–18 April 2019; doi:10.1109/WF-IoT.2019.8767313. [CrossRef]

30. Chang, Y.H.; Ko, C.B. A Study on the Design of Low-Code and No Code Platform for Mobile Application Development. *Int. J. Adv. Smart Converg.* **2017**, *6*, 50–55. [CrossRef]

31. Barricelli, B.R.; Cassano, F.; Fogli, D.; Piccinno, A. End-user development, end-user programming and end-user software engineering: A systematic mapping study. *J. Syst. Softw.* **2019**, *149*, 101–137. [CrossRef]

32. Paternò, F.; Santoro, C. End-User Development for Personalizing Applications, Things, and Robots. *Int. J. Hum. Comput. Stud.* **2019**. [CrossRef]

33. Bau, D.; Gray, J.; Kelleher, C.; Sheldon, J.; Turbak, F. Learnable Programming: Blocks and Beyond. *Commun. ACM* **2017**, *60*, 72–80. [CrossRef]

34. Lifelong Kindergarten Group. Scratch - Imagine, Program, Share, 2019. Available online: https://scratch.mit.edu/ (accessed on 16 October 2019).

35. Armoni, M.; Meerbaum-Salant, O.; Ben-Ari, M. From scratch to "real" programming. *ACM Trans. Comput. Educ. (TOCE)* **2015**, *14*, 25. [CrossRef]

36. Laval, J. End user live programming environment for robotics. *Robot. Autom. Eng. J.* **2018**, *3*. [CrossRef]

37. Massachusetts Institute of Technology. MIT App Inventor, 2019. Available online: https://appinventor.mit.edu/ (accessed on 16 October 2019).

38. David, W.; Abelson, H.; Spertus, E.; Looney, L. *App Inventor: Create Your Own Android Apps*; O'Reilly Media, Inc.: Sebastopol, CA, USA 2015.

39. Leonardi, N.; Manca, M.; Paternò, F.; Santoro, C. Trigger-Action Programming for Personalising Humanoid Robot Behaviour. In Proceedings of the ACM CHI Conference on Human Factors in Computing Systems, Scotland Uk, 4–9 May 2019; pp. 445:1–445:13.

40. Ur, B.; McManus, E.; Yong Ho, M.P.; Littman, M.L. Practical trigger-action programming in the smart home. In Proceedings of the ACM SIGCHI Conference on Human Factors in Computing Systems, Toronto, ON, Canada, 26 April–1 May 2014; pp. 803–812.

41. Lifelong Kindergarten Group. Scratchx, 2019. Available online: https://scratchx.org/ (accessed on 16 October 2019).

42. Massachusetts Institute of Technology. MIT App Inventor + Internet of Things, 2019. Available online: http://iot.appinventor.mit.edu/ (accessed on 16 October 2019).

43. Rizzo, A.; Burresi, G.; Montefoschi, F.; Caporali, M.; Giorgi, R. Making IoT with UDOO. *Interact. Des. Archit. J.* **2016**, *30*, 95–112.

44. Lu, C.H.; Hwang, T.; Hwang, I.S. IoT Inventor: A web-enabled composer for building IoT-enabled reconfigurable agentized services. In Proceedings of the 2016 IEEE International Conference on Consumer Electronics-Taiwan (ICCE-TW), Nantou, Taiwan, 27–29 May 2016; pp. 1–2.

45. Mota, J.M.; Ruiz-Rube, I.; Dodero, J.M.; Arnedillo-Sánchez, I. Augmented reality mobile app development for all. *Comput. Electr. Eng.* **2018**, *65*, 250–260. [CrossRef]

46. Corral, J.M.R.; Ruíz-Rube, I.; Balcells, A.C.; Mota-Macías, J.M.; Morgado-Estévez, A.; Dodero, J.M. A Study on the Suitability of Visual Languages for Non-Expert Robot Programmers. *IEEE Access* **2019**, *7*, 17535–17550. [CrossRef]

47. Rubin, J.; Chisnell, D. *Handbook of Usability Testing: How to Plan, Design and Conduct Effective Tests*; John Wiley & Sons.: Indianapolis, IN, USA, 2008.

48. Krishnamurthi, S.; Fisler, K., Programming Paradigms and Beyond. In *The Cambridge Handbook of Computing Education Research*; Cambridge University Press: Cambridge, UK, 2019.

49. O'Grady, M.J.; Muldoon, C.; Carr, D.; Wan, J.; Kroon, B.; O'Hare, G.M.P. Intelligent Sensing for Citizen Science. *Mob. Netw. Appl.* **2016**, *21*, 375–385. [CrossRef]

Article

Fostering Environmental Awareness with Smart IoT Planters in Campuses

Bernardo Tabuenca [1,*]**, Vicente García-Alcántara** [1]**, Carlos Gilarranz-Casado** [2] **and Samuel Barrado-Aguirre** [3]

[1] Departamento de Sistemas Informáticos, Universidad Politécnica de Madrid, 28031 Madrid, Spain
[2] Departamento de Ingeniería Agroforestal, Universidad Politécnica de Madrid, 28040 Madrid, Spain
[3] Corporación Radiotelevisión Española, Dirección de Tecnología y Sistemas, 28223 Madrid, Spain
* Correspondence: bernardo.tabuenca@upm.es; Tel.: +34-91-06-73552

Received: 14 March 2020; Accepted: 11 April 2020; Published: 15 April 2020

Abstract: The decrease in the cost of sensors during the last years, and the arrival of the 5th generation of mobile technology will greatly benefit Internet of Things (IoT) innovation. Accordingly, the use of IoT in new agronomic practices might be a vital part for improving soil quality, optimising water usage, or improving the environment. Nonetheless, the implementation of IoT systems to foster environmental awareness in educational settings is still unexplored. This work addresses the educational need to train students on how to design complex sensor-based IoT ecosystems. Hence, a Project-Based-Learning approach is followed to explore multidisciplinary learning processes implementing IoT systems that varied in the sensors, actuators, microcontrollers, plants, soils and irrigation system they used. Three different types of planters were implemented, namely, hydroponic system, vertical garden, and rectangular planters. This work presents three key contributions that might help to improve teaching and learning processes. First, a holistic architecture describing how IoT ecosystems can be implemented in higher education settings is presented. Second, the results of an evaluation exploring teamwork performance in multidisciplinary groups is reported. Third, alternative initiatives to promote environmental awareness in educational contexts (based on the lessons learned) are suggested. The results of the evaluation show that multidisciplinary work including students from different expertise areas is highly beneficial for learning as well as on the perception of quality of the work obtained by the whole group. These conclusions rekindle the need to encourage work in multidisciplinary teams to train engineers for Industry 4.0 in Higher Education.

Keywords: computer-based systems; environmental awareness; Industry 4.0; Internet of Things; irrigation systems; planter; project-based-learning; smart learning environments; teamwork

---

1. Introduction

The growth of wireless networks and the proliferation of mobile devices in all contexts of human social life are facilitating the implementation of initiatives in which objects and people are connected. More recently, 5G networks have increased in response speed, transfer speed, data bandwidth, and wireless coverage throughout the whole populated territory on earth [1,2]. In this context, it is key to train engineers with the capabilities to develop systems that combine people, networks, and real-world objects. The Internet of Things (IoT) is widely considered the next step towards a digital society where objects and people are interconnected and interact through communication networks in Smart Cities [3,4].

The decrease in the price of electronic components (i.e., microcontrollers, sensors, and actuators) and their availability to buy them in online markets from home has facilitated the integration of education on microelectronics in schools and universities. Manufacturers of microcontrollers, as an interested party, are putting countless "Starter kits" for sale with low-cost cases to get started in the world of IoT (e.g., Grove Starter Kit for IoT, Adafruit Microsoft Azure IoT starter kit, Arduino starter kit, SparkFun IoT starter kit, CanaKit Raspberry Pi starter kit, Vilros Raspberry Pi 4 complete). Furthermore, social networks are facilitating the sharing of the so-called "know-how" through videos and descriptive documents that help to assemble the components in only some minutes. Nonetheless, the industry not only requires engineers to know how to assemble Lego bricks correctly, but engineers must also understand why bricks are assembled this way. It is essential that universities instil ingenuity in students so that future engineers know how to apply the knowledge in the most effective, efficient and sustainable way towards Industry 4.0.

Industry 4.0 is a concept originated to describe a vision of manufacturing with all its processes interconnected through the IoT. Industry 4.0 (or 4th industrial revolution) consists of digitizing industrial processes, automating tasks by training machines with artificial intelligence, connectivity, and optimizing resources. The combination of Internet technologies and future-oriented technologies in the field of smart objects proves to be a new fundamental paradigm shift in industrial production [5]. The optimization of resources in Industry 4.0 is related to the optimization of production not only at an economic level saving raw materials and energy resources, but also at an ecological level promoting the principles of sustainability to care for the planet. The United Nations 2030 Agenda for Sustainable Development with its 17 Sustainable Development Goals presents a bold and comprehensive framework for development cooperation in the coming years [6]. In this sense, it is important to educate students about the need to implement technological solutions that meet the objectives of the 2030 agenda to ensure that the engineers of the future have more capacity and ecological awareness than their predecessors of the three previous industrial revolutions.

Implementing IoT systems that are economically and ecologically efficient requires a deep knowledge on different disciplines. Creating these systems requires expertise to interconnect the components, but also capability to implement systems that are coherent with a sustainable environment. Nowadays, engineering career Curricula are designed to train students in a specific discipline e.g., Computer Engineering, Agricultural Engineering, Forestry Engineering, Industrial Engineering, Civil Engineering. The duration of the Curricula does not usually allow training more than one discipline, and transversal competences like sustainability or reflective practice on environmental issues are not always explored enough.

In this work, we carry out a study in which future engineers from different disciplines (i.e., Computer Engineering, Agricultural Engineering) work together to implement Smart IoT planters. These Smart IoT planters aim to promote environmental awareness on university campuses. The contribution of this work is threefold: Firstly, a holistic architecture describing the Smart IoT planters implemented throughout a semester is presented. Secondly, the results of an evaluation exploring teamwork performance in multidisciplinary teams are stated. Third, alternative initiatives to promote environmental awareness on campuses from students' perspective are presented.

This article is distributed as follows: in Section 1.1 previous work on Smart IoT planters is explored and the benefits of plants in learning spaces are specified. In Section 1.2 the scientific literature in the context of teamwork in educational settings is examined. In Section 2, the materials and methodology applied in this study are described. In Section 3 the results of the study are presented. Finally, in Section 4 the results are analysed and discussed.

1.1. Smart IoT Planters in Educational Contexts

Indoor plants have countless benefits to human health and well-being [7–11]. The classroom environment can play an important role in students' learning and academic performance [8]. Han [9] performed a study in two different high school classes of which one served as the experimental group and the other as control. Six plants were placed at the back of the classroom. The experimental group had immediately and significantly stronger feelings of preference, comfort, and friendliness as compared to the control group. Also, the experimental group had significantly fewer hours of sick leave and punishment records due to misbehaviour than the control group. Similarly, Fjeld [10] performed a study measuring health and discomfort symptoms. The results concluded 21% lower mean score for health symptoms in classrooms with plants, and a more positive consideration of the classrooms with plants (more beautiful, brighter, and comfortable). Likewise, Khan et al. [11] explored self-reported observations on indoor air quality, aesthetics, and performance. The results show that large majorities reported that the plants improved air quality, increased pleasantness, and helped to improve their performance. This study investigates alternative implementations to install Smart IoT planters as an approach to promote environmental awareness using plants in learning spaces.

Placing IoT systems in plants is not new [12–19]. The availability of sensors and microcontrollers in the online market has facilitated the implementation of systems for monitoring plants, automate irrigation, provide artificial light, or disease detection in plants. Gomez et al. [12] designed a system for monitoring soil moisture, humidity, and ultraviolet radiation in protected crops. Srbinovska [13] presented a wireless sensor network architecture for vegetable greenhouse in order to achieve cultivation and lower management costs monitoring temperature, humidity, and illumination. Similarly, in the context of greenhouses, Lambebo and Haghani [14] designed a wireless sensor network using open source and inexpensive hardware to measure the concentration level of several greenhouse gases. These planters vary on which sensors, microcontrollers, data persistence storage, and actuators are used to grow plants. Recent implementations in higher education contexts are putting special emphasis on the technological challenges faced and on the solutions adopted [20,21], or the impact on students' satisfaction and motivation [22]. However, none of these systems used real plants with IoT systems to foster environmental awareness in educational contexts. Hence, research questions 1 and 2 are formulated as follows:

RQ1. What kind of Smart IoT planters can be implemented to foster environmental awareness in educational contexts?

RQ2. What alternative initiatives can be proposed to promote environmental awareness in the campus?

1.2. Teamwork Towards Multidisciplinary Implementations

Project-Based-Learning (PBL) is a framework for teaching and learning organized activities to create a product that is becoming more popular in the last years [22–25]. Within this framework, students working in groups strive for solutions to complex problems by asking and clarifying questions, debating ideas, making predictions collecting and analysing data, communicating their findings to others and creating artefacts [26]. PBL is based on re-engineered processes that involve people from multiple disciplines to improve and broaden the competence of engineering students. The aim of PBL is to: (a) understand the role of theoretical and real-world discipline-specific knowledge in a multi-disciplinary, collaborative, practical project-centred environment; (b) recognize the relationship of the engineering enterprise to the social/economic/political context of engineering practice and the key role of this context in engineering decisions, and; (c) learn how to participate in and lead multidisciplinary teams to design and build environmentally conscious and high quality facilities faster and more economically [27]. The review of research on PBL concludes that there is a need for more research documenting the effects and effectiveness using this model [28]. In fact, the literature shows several representations to comprehend the effectiveness of learning processes in working teams [29–34], most of which have been produced around the Input-Process-Output (I-P-O) model [32]. *Inputs* set the conditions under which group interaction processes take place [35]. Inputs are variables that can affect teamwork at various levels (e.g., individual, group and environment) both internally (e.g., members' skills, attitudes, personality, group structure, group size) and externally (e.g., level of environmental stress, reward structure). Group interaction *processes* take place when team members interact [35] and indicate how a group is performing [30]. Finally, *outcomes* are criteria to assess the effectiveness of team actions. A recent study [36] draws on the model to conclude that cooperativeness and collaborative behaviour had a positive influence on team cohesiveness, while workload and task complexity had a negative influence on it. Additionally, the study found that team cohesiveness was positively related to perceived learning, satisfaction with teamwork, and expected quality. Likewise, both perceived learning and expected quality predicted satisfaction with teamwork.

Therefore, we investigate the effects and effectiveness of PBL varying the group composition in the context of Higher Education studies [28] considering the instruments and conclusions suggested in [36]. Logically, research questions 3 and 4 are formulated as follows:

RQ3. What are the effects of multidisciplinary teamwork in learning performance?

RQ4. Which are the channels used to communicate in teamwork.

2. Materials and Methods

This study was carried out in the context of the Computer Based Systems module, which runs in the first semester period of the fourth year of the Degree in Computer Engineering. The module is mostly practical, and it is aimed at implementing technological solutions based on IoT for Smart Cities. This year it was decided to suggest the design of Smart IoT planters to automate the irrigation system at the university campus. Implementing an irrigation system requires very specific knowledge and expertise that most computer students do not have. For this reason, teachers decided that students of Computer Based Systems (Computer Engineering (CE)) might join students of Irrigation and Drainage Technology Systems (Agricultural Engineering (AE)) to work collaboratively to implement solutions considering multidisciplinary perspectives. Both CE and AD modules run in parallel along the first semester of the year.

As CE and AE students might broadly differ in their interests towards implementing the planter, a transversal approach was sought. Hence, teachers highlighted that the main objective of the Smart IoT planter should be to raise awareness among students and university employees about the need to cover Sustainable Development Goals (SDGs) [6] specified in modules' syllabuses: Goal 7. Ensure access to affordable, reliable, sustainable and modern energy for all; Goal 11. Make cities and human settlements inclusive, safe, resilient, and sustainable; Goal 12. Ensure sustainable consumption and production patterns; and Goal 13. Take urgent action to combat climate change and its impacts. The modules followed a PBL approach.

2.1. Participants

For this study, all students enrolled (n = 40) in the Computer Based Systems module (CE), and all students enrolled (n = 12) in the Irrigation and Drainage Technology Systems (AE) were invited to participate in the study. Finally, a total of 47 students accepted the consent form and completed the pre-questionnaire (88.71% male, 14.29% female), thereof 35 were CE students and 12 were AE students. The mean age was 24.23 and the standard deviation was 4.08. All sessions were face-to-face, and the attendance was mandatory but not registered.

2.2. Materials

During the first month (September 2019), students were able to plan their project identifying what components they would need, how they would be assembled, and for what purpose. On the one hand, CE students took the lead within their groups by analysing the most suitable sensors, and actuators. The selection and purchase of components was restricted to the stocks in a well-known website specialized in electronic components. On the other hand, AE students took the lead within their groups to decide on which components were required to assemble the irrigation system, the type of soil, and the type of plant. Before making the full purchase, the groups presented a preliminary draft specifying which components were to be used and how they were to be assembled. Professors from both disciplines (CE and AE) validated the proposals, and suggested changes to guarantee the operation of their future IoT projects.

All students had institutional tools to work collaboratively online (i.e., Microsoft Sharepoint, Microsoft Teams, Skype, and email accounts). All students were assisted on-demand by teachers.

2.3. Design of the Case Study

This collaboration between faculties from different disciplines was developed with the aim to promote mixed areas of knowledge (i.e., mechatronics) within the university. This study was carried out and presented to students as an initiative in which they should design IoT systems that are compliant with the SDGs [6].

The composition of the groups varied on the expertise of the students, and the expertise of the teachers. Hence, there were 3 different types of groups:

(1) Type A groups (3 x groups) comprised 4 CE students. Type A groups had support from CE teachers. These groups comprised only members from the CE discipline. These were the groups with the lowest degree of multidisciplinarity.
(2) Type B groups (3 x groups) comprised 4 CE students, and 4 AE students. Type B groups had support from both the CE and the AE teachers. AE students were only assigned to type B groups. These were groups with the highest degree of multidisciplinarity.
(3) Type C groups (3 x groups) comprised only CE students. Type C groups had support from both CE and AE teachers. These were the groups with a medium degree of multidisciplinarity.

Regarding the group formation, teachers presented the 3 types of groups that students might join to work along the semester. Students were assigned to groups as soon as they decided the option that was more convenient for them. Finally, there were 3 groups for each group type. Each group voluntarily chose a leader who was in charge of organizing the communication, and taking care of the materials. All groups collaborated during the semester using the channels they found handier.

CE and AE students were allocated on two different campuses in the city (i.e., Campus Sur, and Ciudad Universitaria). Hence, group leaders organized sporadic and spontaneous meetings to make progress within their projects. Additionally, students were encouraged to use the channels provided by the university to arrange synchronous and/or asynchronous online meetings.

2.4. Meassure Instruments

This study followed the constructs suggested by Gil et al. [36] to measure team work. Reasonably, seven-point Likert scales were used to measure the 8 constructs: (1) Cooperativeness was measured using items 4 from [37]; (2) Collaborative behaviour was measured using 5 items from [38]; (3) Task complexity and (4) Workload were measured using 3 and 4 items respectively from [39]; (5) Team cohesiveness was measured using 6 items from [37,40,41]; (6) Perceived learning was measured using items from [42]; (7) Expected quality was measured using 3 items from [43]; (8) Satisfaction with teamwork was assessed following [41,44]. The forms were completed using an online questionnaire at the end of the course.

Teachers evaluated the work of CE students in a 0 to 10 scale at the end of the course. Students had to prepare a practical demo of the planter, write a technical document describing hardware and software designs, specify how they had covered the SDGs in their implementation, suggest further initiatives to foster environmental awareness in the campus, and openly share the code in a repository. In contrast, AE students were evaluated considering elements of assessment that happened before the start of the study. Hence, AE students' grades were excluded from the evaluation to avoid bias.

2.5. Procedure

This experiment took place between September 2019 and January 2020 in 32 lab sessions of two hours duration (2 sessions per week). To attract the motivation of the students, the origin of urban gardens was contextualized in World War II and linked to its current use in most developed countries where it is becoming an alternative to the consumption of transgenic foods and pesticides. Looking for a closer standpoint, this initiative was motivated in the context of their own campus to make it more sustainable.

The first two weeks of the course comprised the presentation of the module, the introduction to theoretical concepts, and the group formation. Starting the fifth school-week, each group had to deliver a preliminary draft (pre-design of the IoT system) including an outline that specified which components were going to be needed, and how they were going to be interconnected to achieve the specific objectives of their final project. The teachers used the students' proposals to filter, to agree, and to purchase the components based on their expertise and the existing budget. Starting the sixth week, students could start working with the components. At the end of week 16th, groups defended their projects making a demo, writing technical documentation, and answering the questions formulated by the teachers during the evaluation.

2.6. Data Analysis

Questionnaires data and scores were imported from the survey-platform into MS Excel format and then analysed using R Studio (v1.2.1335).

The reliability coefficient (Cronbach's alpha) was calculated to validate the internal consistency of the sample (see Table 1). Nunnally has suggested that score reliability of 0.70 or better is acceptable [45].

Table 1. Overall Means (M), Standard Deviations (SD), and Reliability Coefficient (Cronbach's Alpha).

Scale (7-Point Likert)	M(SD)	Cronbach's α	Sample Item
Perceived learning	4.35(1.68)	0.96	*I am learning to identify the central issues of the subject*
Expected quality	5.75(1.30)	0.94	*I think the work of my team deserves a high mark*
Team cohesiveness	5.67(1.23)	0.89	*Every member in the team fulfils their part*
Workload	2.78(1.38)	0.89	*Teamwork requires a lot of my time*
Satisfaction	5.97(1.08)	0.77	*I enjoy working with my team*
Collaborative behaviour	5.70(0.94)	0.71	*Teamwork is stimulating for me*
Cooperativeness	5.06(0.78)	0.52 *	*I like to work with other people*
Task complexity	4.27(1.25)	0.42 *	*I have undertaken similar tasks in other subjects*

*: Internal consistency ($\alpha \geq 70$).

A Shapiro–Wilk test was conducted to confirm the normal distribution assumption of the sample towards performing an analysis of variance (ANOVA). The ANOVA test was conducted to confirm significant differences among the means obtained.

Finally, a Pearson's correlation analysis was run to determine the relationship between the means obtained (Table 4). Pearson indicates the strength of the linear relationship between two variables for which the values range between $-1 < 0 < 1$. The values closer to 1 (-1) depict a stronger positive (negative) correlation, meaning that the second variable tends to increase (decrease) when the values of the first value are increased and vice versa. The closer the values are to 0, the weaker the correlation is. A p-value less than 0.01 is taken as indicator for significant correlations. We can verbally describe the strength of the correlation using the guide that Evans [42] suggested for the absolute value of r (Strength: 0.00–0.19 "very weak"; 0.20–0.39 "weak"; 0.40–0.59 "moderate"; 0.60–0.79 "strong"; 0.80–1.0 "very strong").

3. Results

The results presented in Section 3.1 address RQ1, Section 3.2 addresses RQ3 and RQ4, whereas RQ2 is addressed in Section 3.3.

3.1. IoT Systems to Foster Environmental Awareness in the Campus

This study investigates what kind of Smart IoT planters can be implemented to foster environmental awareness in educational contexts (RQ1). Finally, the groups implemented 3 different planter types:

1. *Vertical garden*: A wooden pallet placed vertically on a wall with 6 plastic bottles attached to it. The bottles are placed diagonally from the top to the bottom with a change of direction.

When watering the bottle on top using a dripper, the water runs from one bottle to the next one immediately below. The circuit concludes in a small tank for the remaining water. See Figure 1a.

2. *Hydroponic system*: A closed circular circuit created with pipes. The plants are placed in the holes created in the upper part of the pipe so that the base of the plant is in contact with water and nutrients. A pump is used to circulate water from a tank to the rest of the circuit. See Figure 1b.

3. *Rectangular planters*: 6 plastic planters with $50 \times 38 \times 30$ dimensions were installed. A slit was opened at the bottom of the side of the planters to release the remaining water. Figure 1c illustrates how some of the electronic components are embedded.

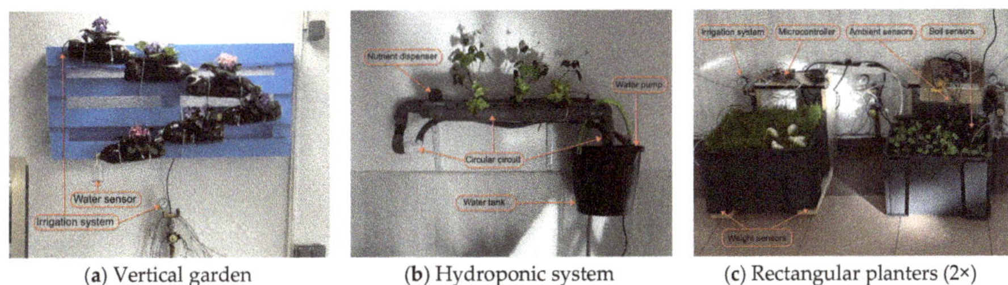

(**a**) Vertical garden   (**b**) Hydroponic system   (**c**) Rectangular planters (2×)

Figure 1. Internet of Things (IoT) planters installed in the campus.

Each working group justified the selected plant considering where it was going to be installed. The groups finally decided on cyca palm, organic grass, lolium perennial, kale, artichoke, and peppermint. Ornamental flowers were planted in the vertical garden, and lettuce was planted in the hydroponic system. The composition of the soil (i.e., substratum and soil) was adapted to the needs of each type of plant.

The planters were equipped with different IoT systems. Sensors, actuators, IoT cloud platform, and irrigation systems included in the IoT systems are illustrated in Figure 2. In the following sections, a holistic approach is represented to describe all the components included by the different planters.

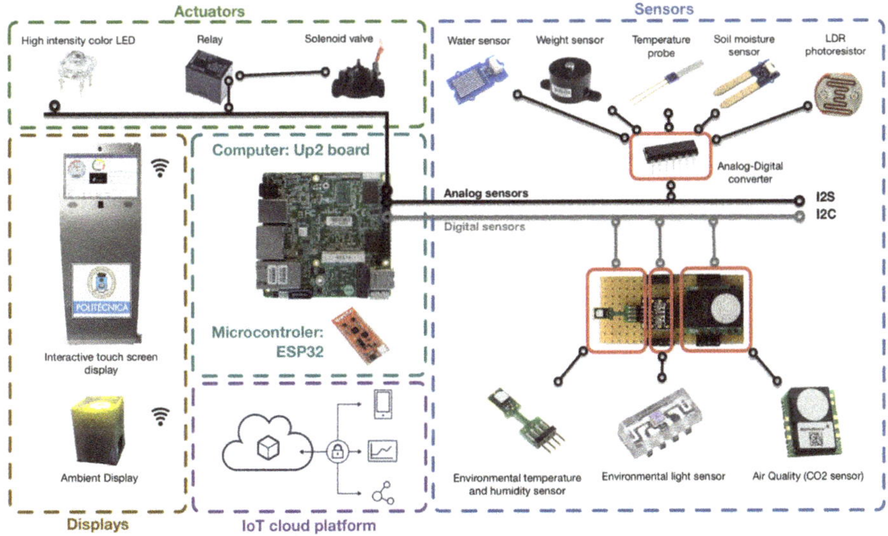

Figure 2. Holistic architecture of the components included in the IoT planters.

3.1.1. Sensors

Students were able to investigate which variables they needed to manage and therefore what sensors they wanted to install in their planter. The groups selected both analog and digital sensors. Analog sensors return voltage as an output whereas digital sensors return digital values. These were the sensors installed in the planters (See Figure 2):

- *Water sensor*: an analog sensor which returns 0 value if no water is detected, and a higher value when water is detected. The vertical garden was designed placing the water sensor within a small tank attached to the bottleneck of the last bottle. This tank contains the excess water. Similarly, all rectangular planters have a slit in the bottom back side where excess water can escape to a plate when the pot overflows. The water sensor was placed on the plate so that whenever there was a drop on its grid, the irrigation system was automatically stopped.
- *Weight sensor:* an analog sensor that varies the output voltage depending on the mass on it. This sensor is placed at the base of the planter of rectangular planters to keep track of the weight. The groups installed this sensor with two different purposes: (1) Identify when to stop the irrigation. Knowing how much weight the pot has before starting to water, and how much weight the pot has when the water begins to overflow, the students were able to design a system to stop watering.; (2) Manage the evolution of the plant mass of the planter. The plant grows inside and outside the pot as time passes. This sensor allows you to know precisely what plant mass the pot has. This data is relevant to determine the amount of nutrients needed, and to determine the moment when to relocate the plant to a larger pot.
- *Temperature probe*: an analog sensor using a resistance that varies the output voltage between 0 and 1000 depending on the inner temperature. The probe is installed inside the land to explore the temperature of the plant at different depths.
- *Soil moisture sensor*: an analog sensor that measures the volumetric water content in soil. Measuring soil moisture is important to manage the irrigation system more efficiently. This sensor measures the humidity of the soil of the plant positioning the two legs inside the ground. The students proposed the use of this sensor in 2 different ways: (1) To control humidity horizontally. Students placed the sensor on the surface of the land to determine the humidity of the soil at different distances from the drip system (or plant stem); (2) to control humidity vertically. Students placed the sensor making a slit in the side of the pot at different depths to measure the evolution of the wet bulb.
- *Light Dependent Resistor* (LDR): an analog sensor whose resistance varies depending on the amount of light falling on its surface. These resistors are often used in circuits where it is required to sense the presence of light. The groups used this sensor to artificially adapt the light of the plant to make the photosynthesis, when natural light was not appropriate.
- *Environmental temperature*: a digital sensor that returns two decimal values reporting Celsius degrees. The temperature in campus corridors usually fluctuates depending on the time of day, the angle of the light, if windows are open or closed, number of students around, or if the heating system is on or off. Likewise, there are plants that are more sensitive than others to temperature fluctuations. The ambient temperature sensor allowed to monitor how the temperature varied throughout the day and to provide suitable feedback depending on the type of plant being cultivated.
- *Environmental humidity*: a digital sensor that returns two decimal values reporting percentage of humidity. Moisture is important so that photosynthesis is possible. Likewise, plants should not lose too much water from their leaves. The humidity sensor allows to monitor how the humidity varied and to adapt the feedback to the user based on the need to humidify the plant. Some students suggested installing an air humidifier as actuator in further implementations.
- *Environmental light*: a digital sensor, which returns four decimal values between 0 and 2400 reporting the existing light intensity measured in lux (unit of measure for the amount of light

received by the sensor). Similar to the LDR, it is used to provide additional artificial light if the natural one is not enough.

- *CO_2-Air quality*: a digital sensor that returns four decimal values between 450 and 2000 ppm (parts per million). Air quality is measured on the basis of the CO_2 ppm number and volatile organic compounds coexisting in the air. The students implemented this sensor to alert users when the corridor is saturated with CO_2 and it is necessary to open the windows.

3.1.2. Actuators

- *Relay*. The relay is a switch controlled by an electrical circuit by means of a coil and an electromagnet to open or close the electro valve.
- *Solenoid valve* (electro valve). This valve controls the passage of the irrigation water through the pipe. The valve is moved by a solenoid coil, and has only two positions: open or closed.
- *High intensity colour LED*. The students used sets of LEDs for two different functionalities: (1) produce artificial light to facilitate photosynthesis of the plant; (2) provide feedback to the user in real time on how the irrigation system is working. e.g., Group #1 implemented the following policy: LED lights blue when the plant is watering; the LED turns red when the AQ sensor reported over 1000 ppm of CO_2; the LED turns green when the sensors of the plant return optimal values.
- *Ambient displays and feedback tools*. Students used different interfaces to show the values returned by the sensors:

  (a) Ambient display: The PRISMA is an environmental display to support learning scenarios [46]. See Figure 3a. The PRISMA can display information with its 24 LED ring, 8 × 8 LED matrix, and a liquid crystal display. This display was made available to students so they could configure it based on their interaction needs. E.g. Group #3 configured the PRISMA to provide a range of colours between blue and yellow derived from the humidity returned by the sensor.

  (b) Interactive touch screen display. This interactive display was installed to present real-time information from all planters making sensors data visible with visual metaphors. The main screen includes a menu where the user can select which planter to explore in detail. See Figure 2 (left).

  (c) Mobile messaging system. Group #5 configured the IoT system to send alerts by means of Telegram instant messaging app, when specific events occur. Figure 3b shows some examples of the configured alerts: "The plant has not enough light", "Congratulations! The plant is growing under the best conditions", "Security alert! There is no Internet connectivity. Check the planter". The system also notifies when the irrigation has started and finished.

3.1.3. Computer and Microcontroller

The board (or microcontroller) is the main component connecting the rest of the subsystems. Its role is to receive the captured data, processing the data, and send orders to actuators to maintain the plant in the best conditions. Likewise, the processor sends data to the IoT platform where it is stored and monitored according to consistent rules. The system is continuously active to periodically read, validate, and write the value of the sensors. The IoT systems implemented included both Up_2 board (computer) and/or ESP32 (microcontroller).

- *UP^2 board*. The UP Squared board is an x86 maker board based on the Intel. The UP boards are used in IoT applications, industrial automation, or digital signage. This board is equipped with an Intel Celeron N3550 and Intel Pentium N4200 System on Chip (SoC), 40 pins, 8 GB RAM, Ethernet, HDMI, and USB connectors. This case study was carried out along the semester of the

Computer Based Systems module. Hence, students were urged to implement their IoT systems using this board.

- *ESP32 microcontroller.* ESP32 is a series of low-cost, low-power SoC microcontrollers with integrated Wi-Fi and dual-mode Bluetooth. The ESP32 employs a Tensilica Xtensa LX6 microprocessor. ESP32 includes built-in antenna switches, power amplifier, low-noise receives amplifier, filters, and power-management modules. In this case study, the most advantageous groups were able to adapt the processing capacity of the system and replace the board with a microcontroller.

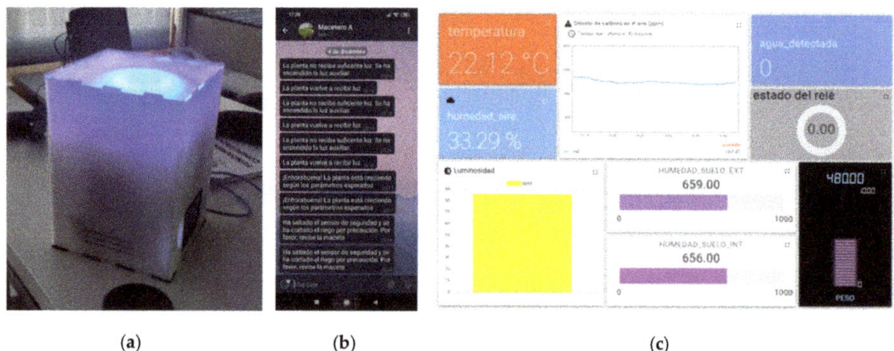

(a)   (b)   (c)

Figure 3. Feedback services configured: (**a**) Prisma, a visual feedback display; (**b**) Telegram messenger to receive alerts; (**c**) IoT cloud platform. Thingsboard.io desktop dashboard.

3.1.4. IoT Cloud Platform

In the initial phase, students should agree on which IoT cloud platform would be used to persist and monitor data from planters. A brainstorming session was organized to explore and test existing IoT platforms. The following features were considered to take the decision: REST API, authentication type, protocols for data collection (i.e., MQTT, HTTP, CoAP), and analytics provided. The following IoT platforms were considered: Azure IoT, DeviceHive, Kaa IoT Platform, Mainflux, SiteWhere, Thingsboard.io, Thinger.io ThingSpeak, SSo2, and Zetta. Finally, Thingsboard.io received more votes from the students at the end of the brainstorming session.

All groups created a profile in Thingsboard.io to adapt the IoT platform to the specific requirements of each planter. These were the features used by the groups:

- *Application Programming Interface* (API). All groups used the Rest API to remotely store the data in the cloud via HTTP or MQTT protocols.
- *Rule engine.* Students were able to configure specific rules to validate the data and consistently perform specific actions. e.g., Group #5 configured the platform to broadcast mobile messages alerting the user via Telegram when precise events occurred.
- *Data persistence.* Trial profiles created in the IoT platform had restrictions regarding the duration of the data persistence in the cloud. Hence, some groups were able to configure the system to backup the data in a local database to keep long-term data.
- *Visualization dashboard.* The IoT dashboard is a key HMI (Human-Machine Interface) component to organize and present digital information from the physical world into a simply understood display on a computer or mobile. Hence, students were able to interpret the information stored in the IoT platform using different interfaces depending on the sensor they were able to configure (See Figure 3c).

3.1.5. Irrigation System

Agricultural Engineering students and teachers were responsible for designing and installing the automatic irrigation system. The installation involved the setup of planters, motorized valves, droppers, water pipes, water counters, water filters, and pressure switches. This subsystem is formed by a relay that controls the opening and closing of a solenoid valve, which allows or not the passage of water. The relay receives a direct order from the computer/microcontroller configured by CE students. CE and AE teachers provided on-demand support but also regularly reviewed the progress performed by each group.

3.2. Mutlidisciplinary Teamwork on IoT

This study is aimed at exploring the effects of multidisciplinary teamwork in learning performance (RQ3). Hence, we explored the extent to which teamwork subscales can vary based on the multidisciplinarity in the composition of the work groups. The scores obtained demonstrated adequate internal consistency for 6 out of 8 scales (see Table 1). Values for Cronbach's alpha ranged from 0.71 to 0.96 revealing sufficient score reliability for "perceived learning", "expected quality", "team cohesiveness", "workload", "satisfaction" and "collaborative behavior". Nevertheless, values for Cronbach's alpha ranged from 0.42 to 0.52 revealing insufficient score reliability for "cooperativeness" and "task complexity", and they were consistently discarded for the rest of the analysis.

Means and standard deviation were calculated taking into account the composition of the groups. The results illustrated in Table 2 show that by taking together all the values of the teamwork scale in a range of 1 to 7, group B obtained an average rating slightly higher than group C. On the contrary, group A obtained a rating of 0.59 points lower than group B. Looking at the subscales individually, the results concluded that group B obtained the highest scores for "collaborative behaviour", "satisfaction", "team cohesiveness", and "perceived learning". On the other hand, group C obtained the highest scores in "expected quality" and "perceived learning". On the contrary, group A obtained the lowest scores in all subscales.

Table 2. Teamwork means and standard deviations by group composition.

Scales/Subscales	Group A	Group B	Group C
	M(SD) N = 15	M(SD) N = 12	M(SD) N = 12
Overall teamwork	4.67(0.88)	5.26(1.01)	5.24(0.61)
Collaborative behaviour	5.69(0.91)	5.80(1.10)	5.58(0.85)
Satisfaction	5.55(1.32)	6.47(0.89)	5.95(0.76)
Team cohesiveness	5.28(1.33)	5.93(1.28)	5.84(1.04)
Expected quality	5.12(1.32)	5.97(1.43)	6.27(0.75)
Perceived learning	4.00(1.63)	4.65(1.79)	4.47(1.71)
Workload	2.38(0.69)	2.72(1.63)	3.31(1.63)
Grades	7.49(1.63)	9.09(0.69)	8.65(0.63)

With regard to the marks obtained by the students in the final evaluation on a scale of 0 to 10 (10 being the best score), the results showed that group B obtained the upper average grade, followed by group C. The group A scored 1.60 points lower than group B, and 1.16 points lower than group C.

The results obtained in the Shapiro–Wilk test (p-value = 0.00057) confirmed the normal distribution of the overall teamwork data sample. Exploring the teamwork subscales independently, p-values lower than 0.05 and the observations of the Q-Q plots confirm that "Collaborative behaviour", "Satisfaction", "Team cohesiveness", "Expected quality", and "Workload" samples are normally distributed. However, the p-values obtained in the "perceived learning" sample deviate from normality. Hence, "perceived learning" was consistently discarded in the ANOVA test.

An ANOVA test was performed to identify significant differences between the mean values (Table 3). On the one hand, the test resulted in significant values for the grades obtained in the evaluation (See grades in Figure 4a). On the other hand, the test resulted in Pr(>F) = 0.17 (which is slightly higher to the coefficient of significance 0.1) for the overall teamwork scales, and consequently non-significant values for the overall teamwork scales. Exploring the values obtained in the subscales, the ANOVA test resulted in significant values for "expected quality" (Figure 4b).

Table 3. Analysis of Variance ANOVA. Significance Pr(>F) > 0.1.

Scales	Sum of Squares	df	Mean Square	F	Pr(>F)
Teamwork	**2.27**	**2**	**1.35**	**1.84**	**0.17**
Collaborative behaviour	0.26	2	0.13	0.14	0.86
Workload	5.25	2	2.62	1.41	0.25
Team cohesiveness	3.11	2	1.55	1.02	0.37
Expected quality	8.56	2	4.28	2.81	0.07
Satisfaction	5.04	2	2.52	2.31	0.11
Grades	**17.14**	**2**	**8.57**	**6.73**	**0.003 ****

Pr(<F) Significance codes: 0 '***' 0.001 '**' 0.01 '*' 0.05 '.' 0.1 ' ' 1.

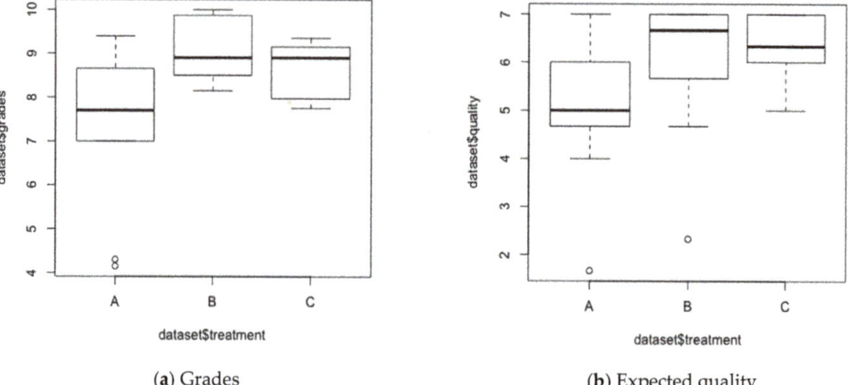

(a) Grades   (b) Expected quality

Figure 4. Boxplots contrasting group composition.

This study aimed at exploring the significance of the relationship between the means obtained in teamwork subscales and the grades. Additionally, we investigated the relationships within teamwork subscales (Table 4). We anticipated that learning performance (grades) would be positively correlated with teamwork subscales. Contrary to our expectations, though, the results of the analysis do not depict a significant correlation between them. Additionally, we aimed at exploring potential correlations within teamwork subscales. The results from the correlation analysis show that there is a significant very strong positive correlation between "cohesiveness" and "satisfaction". Similarly, "cohesiveness" has a strong positive correlation with "expected quality", "perceived learning", and "collaboration". Likewise, "expected quality" has a strong positive correlation with "satisfaction".

Frequently Used Channels to Communicate

In this case study, we aimed at exploring the most frequently used channels to communicate while working in groups (RQ4). Students were assigned an optional task in which they could report once a week (using an online form) what channels they had used to communicate among team colleagues.

There were 272 reports from 47 different students along the semester (Table 5). There was a mean of 5.78 reports by student. Whatsapp, Telegram and Teams were the most frequently used channels.

Table 4. Pearson's correlation analysis (* Correlation significance < 0.01).

r	Grades	Collaborat.	Workload	Cohesive.	Learning	Quality	Satisfact.
Grades	1						
Collaboration	−0.03	1					
Workload	0.09	−0.12	1				
Cohesiveness	0.09	0.61 *	0.12	1			
Learning	0.09	0.59 *	−0.01	0.62 *	1		
Quality	0.34	0.40	0.12	0.66 *	0.35	1	
Satisfaction	0.31	0.56 *	−0.02	0.81 *	0.47	0.67 *	1

Table 5. Channels used to communicate among team colleagues. Frequency of usage.

	Never 1	Rarely 2	Occasionally 3	Frequently 4	Very Frequently 5	M(SD)
	%(n)	%(n)	%(n)	%(n)	%(n)	
WhatsApp	14.34 (39)	5.51 (15)	12.50(34)	22.43(61)	45.22(123)	3.79(1.43)
Telegram	47.06 (128)	7.35 (20)	5.15(14)	5.15(14)	35.29(96)	2.74(1.83)
eMail	53.68 (146)	20.59 (56)	15.81(43)	6.62(18)	3.31(9)	1.85(1.11)
Teams	74.63 (203)	13.97 (38)	9.19(25)	2.21(6)	0(0)	1.39(0.75)
Skype	87.13 (237)	9.56 (26)	2.21(6)	0.74(2)	0.37(1)	1.18(0.53)
Facebook	99.26 (270)	0.74 (2)	0(0)	0(0)	0(0)	1.01(0.08)
Instagram	99.26 (270)	0.74 (2)	0(0)	0(0)	0(0)	1.01(0.08)

3.3. Educational Initiatives to Promote Environmental Awareness in the Campus

Based on the lessons learned along the semester in which the IoT planters were developed, students were encouraged to suggest alternative educational initiatives to promote environmental awareness in the campus. These were the most relevant actions:

There were different students suggesting interactive systems to improve IoT planters. Student #612 suggested developing a mobile app inspired on the Tamagotchi metaphor to take care of the IoT planters. The Tamagotchi would simulate the real IoT planter where participants could vary feeding substances or watering frequency to explore how the real planter would behave. Likewise, she suggested that the Tamagotchi might learn from the experience enabling using artificial intelligent algorithms. Following a similar approach, student #311 proposed to show a status summary of all planters in a large ambient display at the campus using Tamagotchi metaphor. Particularly, ambient displays were pinpointed as a key channel to promote environmental awareness in the campus. Student #313 proposed to show graphs illustrating water expenses by student-day, student-semester, etc., contrasting the number of litters with the ones contained in a glass of water, in a swimming pool, or the ones used by the IoT planters. Student #307 proposed to explore the operation of the planter to contrast the current energy supply with a self-supply system based on solar panels. Student #112 put forward using Twitter as feed to post actions done on the planters so anyone could track how IoT planters take care of the plants. Additionally, he also suggested posting environmental variables collected by the sensor making the data visible illustrations that catch the attention of users. These social interactions might help students understand the real environmental conditions in the campus and consequently to take action to reduce pollution in a local area. Similarly, student #311 suggested creating a hashtag (e.g., #smartIoTplanters) to track the evolutions of the plants across social networks. Student #307 proposed that voluntary students could take responsibility to take care of the IoT planters along the semester. He encouraged them to create an internal competition in which the students who were able to configure the planters with the most suited parameters according to plant, environmental conditions, but also considered introducing machine learning algorithms would save extra money for credits to study related modules at the university. Thanks to advances in technology and machine learning, chatbots have become more popular than ever in recent years. Student #912 proposed using a chatbot (text or voice chatbot) to automate remote actions regarding the planters using commands. He named it *ChatbIoT* and these would be a sample command: (1)–User: "Hi Planter 3!, can you please water the lettuces at 18:00".–ChatbIoT: "For how long?". User: "Five minutes please". ChatbIoT: "Got it!. I will water the lettuces for 5 minutes at 18:00. Is that correct?". User: "Yes, thank you Planter 3!". ChatbIoT: "Okay, I will do so. Hasta la vista, baby.". The ChatbIoT might also help to remember recurrent actions performed on the planter (e.g., when was the last time that the hydroponic garden was fed).

Many students suggested gamification activities to raise awareness on environmental issues in the campus. Hence, different actions and rewards were considered to implement these games:

(a) Actions: Student #311 suggested that some students should be rewarded whenever they maintain the plants and take care of the IoT systems once the module finished. He also suggested creating a campaign in which students could create 1 min video pills to denounce issues happening in the campus to boost the impact in social networks. Consequently, the most impacting denounces would be rewarded. Student #111 suggested reducing paper waste rewarding students who upload their class notes into Moodle. Student #611 believed that the university could create a social game suggesting 1 individual achievable challenge for each of the of the 17 sustainable development goals. Student #612 reported that loyalty recycling in the campus should be rewarded (e.g., paper, batteries, plastics). Student #812 suggested remunerating car sharing and the use of bikes to commute to the campus.

(b) Rewards: Student #111 would reward the most environmentally loyal students with scholarships, ticket bonuses (to exchange in cafeteria, coffee machines, printing machines, or events organized by the university), or grant priority to book the best learning spaces in the library (e.g., quieter, with more light). Student #114 would reward these students providing with priority to select their preferred time slots to attend to the modules distributed in alternative schedules along the day/week. Student #311 suggested providing visibility to the most active students presenting an updated ranking in visual displays.

Different students suggested organizing interactive workshops in the course of the semester. For example, Student #107 suggested creating a physical open space to show the potential benefits of using renewable energy in the campus. Student #212 proposed to create a mailbox where students could contribute with ideas to make the campus more sustainable. Student #313 argued on the importance of healthy habits, foods and plants, and considered that regular workshops should be organised to make students aware of its benefits in long term. Similarly, student #607 urged to practically explore the composition, substances, and pesticides in cultivation soils using sensors, and to promote the understanding on bio food. Student #411 recommended organizing a hackathon to develop IoT software/hardware to reward the most eco-efficient developments (i.e., computation, energy consumption).

There were different students suggesting the creation of associations to promote achievable actions to support the sustainable development goals. Student #107 suggested that creating an association would have power enough to push a "Campus without plastics "action. Student #111 believed that an association should be able to promote the replacement of toilets to save water, i.e., "a presence sensor might count the number of litters wasted every day. A monthly summary could be presented visual displays at the campus". Student #312 recommended promoting situational awareness organizing excursions to recycling plants, or places affected by pollution. Additionally, she proposed to create a compost area in the campus garden. Last but not least, student #307 suggested the use of alternative media channels such as Radio Campus Sur to disseminate good practices in the context of environmental awareness.

4. Discussion and Conclusions

The rapid spread of IoT technologies has triggered the educational challenge of training future engineers to be able to design complex sensor-based ecosystems. In Industry 4.0, it is essential to educate students about the need to implement technological solutions that meet the objectives of the 2030 agenda, and to ensure that the engineers of the future consider ecological issues in their implementations.

This work presents the results of a case study in which students of Agricultural Engineer and Computer Engineer were assigned the task to create IoT systems to promote environmental awareness in the context of a university campus.

Multidisciplinary groups followed the PBL methodology to investigate IoT solutions that varied in the sensors, actuators, microcontrollers, plants, soils and irrigation system they used (RQ1). Section 3.1 elaborates on the three types of planters implemented, namely, *hydroponic system*, *vertical garden*, and *rectangular planters*. Sensors were configured considering the singularities (e.g., irrigation system) of each planter, but also reflecting on the particular care required by each plant. The holistic architecture of the components represented in Figure 2, shows that the implementation of the IoT planters covered up to 9 sensors (thereof 5 were analog, and 4 were digital). The working groups had the ability to engineer functionally different systems using the same sensors (Section 3.1.1). Different feedback systems were assembled to foster understanding on environmental issues in the campus i.e., LCD lights, ambient displays [46], mobile applications, and desktop-oriented dashboards (Section 3.1.2).

The overall architecture implemented in the case study shows that all groups were consistent featuring a three-layered architecture (Figure 5):

1. Input layer. This layer includes the sensors collecting measurements regularly, and the processing of the data done by the microcontroller/computer (Section 3.1.3). Data is sent to the process layer via MQTT protocol.

2. Process layer. Data is stored in a database included in the IoT cloud platform. Rules might be configured according to the specific requirements of each planter (e.g., send a mobile message via Telegram alerting the user when the *water sensor* detects spilled water). This layer includes input/output API with endpoints to store/request data in/from the IoT platform via HTTP protocol.

3. Output layer. Actions configured to be accomplished by the actuators (i.e., alerts, enable irrigation system, disable artificial lighting system). It comprises both commands towards IoT planter maintenance (Section 3.1.5), and feedback information for third party clients using the API interface.

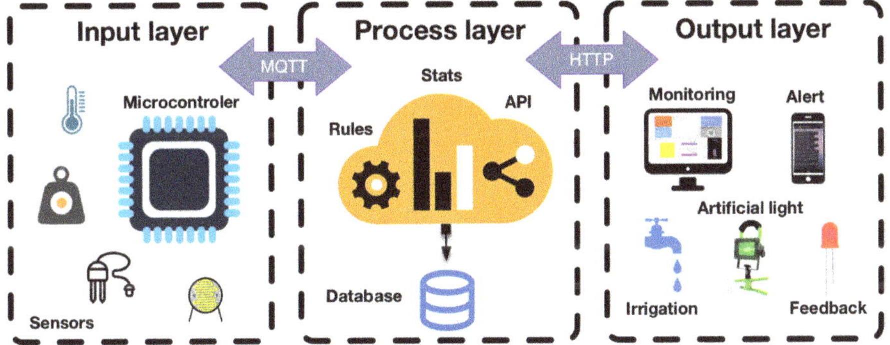

Figure 5. Frequently implemented three-layered architecture in IoT ecosystems.

The architecture described in this section is frequently implemented in different engineering areas [47,48]. Therefore, IoT cloud platforms are providing improved services to facilitate a seamless integration of IoT ecosystems (Section 3.1.4). Due to the scalable nature of the proposed architecture, the ecosystem is easy to extend, and to adapt towards further implementations in educational contexts.

This study achieved the objective of fostering environmental awareness on the campus by implementing systems that addressed the following important issues:

- Optimization the irrigation systems using weight sensors, water sensors, and warning systems. Students implemented IoT planters that open/closed the electro-valve based on real time data, and the particular conditions required by each plant. Likewise, suitable alerts were configured to minimise energy consumption;

- On-site real-time feedback. The IoT planters developed included on-site feedback that draw attention to environmental and consumption variables, promoting the discussion in students walking next to the planters. The LEDs system provided visual feedback specifying when the irrigation stopped/started, when the CO_2 level is over the configured limit, or, when the environmental sensors return optimal values for the plant. The ambient display was configured to make the soil variables visible (i.e., humidity, temperature, weight), transforming the gradient into a colour scale (Figure 3a);

- Online real-time feedback. The systems implemented included different software clients that obtained and reported real-time data. One example was the touchscreen display, which showed detailed graphics on the evolution of the variables of each planter (Figure 3c). Another example was the mobile chatbot, which alerts and traces irrigation schedules (Figure 3b).

Moreover, students were able to suggest alternative educational initiatives to raise awareness about environment issues in the campus (RQ2). The reported results show that the experience with the smart IoT planters facilitated students to envision multiple ingenious strategies to improve IoT planters towards educational purposes (Section 3.3). Different proposals suggested the inclusion of interactive visual and acoustic displays to increase the impact on students and employees of campus. The proposals included the use of social networks, and radio media to share the data collected by the sensors, and the creation of learning objects (in videos, text) to disseminate good practices in the use of water and energy or alert about the pollution in the campus. Furthermore, students suggested solutions based on gamification strategies, designing ingenious reward policies for active users raising environmental awareness. In this context, students were able to identify different achievable activities that could be promoted within any campus. These initiatives imply an important base of knowledge towards implementing further actions to promote environmental awareness in educational contexts.

Multidisciplinary education is key to tackle complex projects covering different areas of knowledge. The implementation of smart IoT planters in educational contexts demanded technical expertise to assemble hardware components and programing software interfaces. Nonetheless, it also required specific expertise on agronomics to select convenient plants, soils, planters, and, to install suitable irrigation systems. In this case study, we wanted to investigate how multidisciplinarity working in groups might impact learning performance and the expected quality of the outcomes (RQ3). The results presented in Section 3.2 show that multidisciplinary work among students from different areas of knowledge is ostensibly beneficial for learning. The groups with the highest degree of multidisciplinarity obtained higher *grades* (Figure 4a). Likewise, a similar effect was observed for the perception of *expected quality* of the work (Figure 4b). Exploring the subscales comprising teamwork, the results of the analysis show that the perception of group cohesion (*team cohesiveness*) and *satisfaction* to work in groups are strongly correlated (Table 4). These findings are consistent with previous research concluding that *expected quality* predicts satisfaction with teamwork [36]. These results rekindle the need to promote work in multidisciplinary groups from different areas of expertise to achieve a deeper knowledge, and to create functionally efficient IoT systems in Higher Education contexts.

Work-in-groups implied both face-to-face and remote to collaboration among group members. Hence, students could freely use different channels of communication to coordinate their activities. On the one hand, they had a default institutional platform (based on Microsoft Teams), which featured synchronous and asynchronous messaging system. Alternatively, they might use any personal messaging applications or social networks. The results collected during the course show that students were more active using personal messaging application from their own mobile devices (Table 5): i.e., WhatsApp, Telegram. However, students were reluctant to use social networks for academic purposes: i.e., Facebook, Instagram (RQ4). These results must be interpreted together with recent conclusions on mobile learning showing that the new generation of teachers would be willing to use personal devices to guide students in educational contexts [49].

The complexity of this study lied in the implementation of a PBL activity across a full semester synchronizing students from two different engineering degrees (Computer Engineering and Agricultural Engineering), campuses, and modules (Computer based systems, and Irrigation and Drainage Technology Systems). This approach is especially difficult to implement since engineering degree syllabi are usually tightly restricted to specific areas of expertise. The results reported in this work represent an important technical knowledge base towards the implementation of IoT ecosystems in CE and AE educational contexts. Likewise, the conclusions of this evaluation provide evidence of the need to encourage work in multidisciplinary teams to train engineers towards Industry 4.0.

Further research should explore alternative associations between multidisciplinary groups to define suitable IoT architectures towards suggesting supplementary learning/teaching paths for industrial, civil, naval, aerospace, or forestry engineering studies.

Supplementary Materials: The following are available online at https://vimeo.com/392389154, Video eUrbanGarden: Campus Sur. UPM.

Author Contributions: Conceptualization, B.T., V.G.-A., C.G.-C., and S.B.-A.; methodology, B.T.; software, B.T. and S.B.-A.; validation, V.G.-A. and C.G.-C.; formal analysis, B.T.; investigation, B.T.; resources, V.G.-A. and C.G.-C.; data curation, B.T.; writing—original draft preparation, B.T.; writing—review and editing, B.T., V.G.-A., C.G.-C., and S.B.-A.; visualization, B.T.; supervision, B.T., V.G.-A., C.G.-C., and S.B.-A.; project administration, V.G.-A. and C.G.-C.; funding acquisition, B.T., V.G.-A. and C.G.-C. All authors have read and agreed to the published version of the manuscript.

Funding: This work received partial support from the Colegio Oficial de Ingenieros Técnicos Agrícolas de Centro, and the Escuela Técnica Superior de Ingeniería de Sistemas Informáticos. Likewise, this work has been co-funded by the Madrid Regional Government, through the project e-Madrid-CM (S2018/TCS-4307).

Acknowledgments: The authors would like to thank Pedro Martínez Jorde for his invaluable support and advice assembling the components. Likewise, the authors would like to thank Trinidad González González for her expert assistance checking English language and style.

Conflicts of Interest: The authors declare no conflict of interest.

References

1. Ge, X.; Zhou, R.; Li, Q. 5G NFV-Based Tactile Internet for Mission-Critical IoT Services. *IEEE Internet Things J.* **2019**, *14*. [CrossRef]
2. Chettri, L.; Bera, R. A Comprehensive Survey on Internet of Things (IoT) toward 5G Wireless Systems. *IEEE Internet Things J.* **2020**, *7*, 16–32. [CrossRef]
3. Kanwal, M.; Malik, A.W.; Rahman, A.U.; Mahmood, I.; Shahzad, M. Sustainable Vehicle-Assisted Edge Computing for Big Data Migration in Smart Cities. *IEEE Internet Things J.* **2020**, *7*, 1857–1871. [CrossRef]
4. Zanella, A.; Bui, N.; Castellani, A.; Vangelista, L.; Zorzi, M. Internet of Things for Smart Cities. *IEEE Internet Things J.* **2014**, *1*, 22–32. [CrossRef]
5. Lasi, H.; Fettke, P.; Kemper, H.G.; Feld, T.; Hoffmann, M. Industry 4.0. *Bus. Inf. Syst. Eng.* **2014**, *6*, 239–242. [CrossRef]
6. Colglazier, W. Sustainable development agenda: 2030. *Science* **2015**, *349*, 1048–1050. [CrossRef]
7. Bringslimark, T.; Hartig, T.; Patil, G.G. The psychological benefits of indoor plants: A critical review of the experimental literature. *J. Environ. Psychol.* **2009**, *29*, 422–433. [CrossRef]
8. Doxey, J.S.; Waliczek, T.M.; Zajicek, J.M. The Impact of Interior Plants in University Classrooms on Student Course Performance and on Student Perceptions of the Course and Instructor. *HortScience* **2009**, *44*, 384–391. [CrossRef]

9. Han, K.-T. Influence of Limitedly Visible Leafy Indoor Plants on the Psychology, Behavior, and Health of Students at a Junior High School in Taiwan. *Environ. Behav.* **2009**, *41*, 658–692. [CrossRef]

10. Fjeld, T. The Effect of Interior Planting on Workers and School Children. *Int. Hum. Issues Hortic.* **2000**, *10*, 46–52.

11. Khan, A.R.; Younis, A.; Riaz, A.; Abbas, M.M. Effect of Interior Plantscaping on Indoor Academic Environment. *J. Agric. Res.* **2005**, *43*, 235–242.

12. Gómez, J.; Castaño, S.; Mercado, T.; Garcia, J.; Fernandez, A. Internet of Things (IoT) system for the monitoring of protected crops. *Rev. Ing. e Innovación* **2018**, *5*, 24–31.

13. Srbinovska, M.; Gavrovski, C.; Dimcev, V.; Krkoleva, A.; Borozan, V. Environmental parameters monitoring in precision agriculture using wireless sensor networks. *J. Clean. Prod.* **2015**, *88*, 297–307. [CrossRef]

14. Lambebo, A.; Haghani, S. A Wireless Sensor Network for Environmental Monitoring of Greenhouse Gases. In Proceedings of the ASEE 2014 Zone I Conference, Bridgpeort, CT, USA, 3–5 April 2014.

15. Karim, F.; Karim, F.; Frihida, A. Monitoring system using web of things in precision agriculture. *Procedia Comput. Sci.* **2017**, *110*, 402–409. [CrossRef]

16. Kim, N.S.; Lee, K.; Ryu, J.H. Study on IoT based wild vegetation community ecological monitoring system. In Proceedings of the International Conference on Ubiquitous and Future Networks (ICUFN), Sapporo, Japan, 7–10 July 2015.

17. Hinojosa-Pinto, S. Diseño de una Arquitectura IoT para el Control de Sistemas Hidropónicos. Final. Ph.D. Thesis, Universidad Politécnic de Valencia, Valencia, Spain, 2019. Available online: https://riunet.upv.es/handle/10251/127335 (accessed on 13 April 2020).

18. Anaya-Isaza, A.; Peluffo-Ordoñez, D.H.; Ivan-Rios, J.; Castro-Silva, J.A.; Ruiz, D.A.; Llanos, L.H. Sistema de Riego Basado En La Internet De Las Cosas (IoT). In Proceedings of the Jornadas Internacionales FICA, Ibarra, Ecuador, 14–15 November 2016; Volume 2016, pp. 1–9.

19. Escalas-Rodríguez, G. *Diseño y Desarrollo de un Prototipo de Riego Automático Controlado con Raspberry Pi y Arduino*; Universitat Politècnica de Catalunya: Barcelona, Spain, 2015.

20. Alvarez-Campana, M.; López, G.; Vázquez, E.; Villagrá, V.; Berrocal, J. Smart CEI Moncloa: An IoT-based Platform for People Flow and Environmental Monitoring on a Smart University Campus. *Sensors* **2017**, *17*, 2856. [CrossRef]

21. Altares-López, S.; Barrado-Aguirre, S.; Loizu-Cisquella, M.; Tabuenca, B.; García-Alcántara, V.; Rubio-Caro, J.-M.; Gilarranz-Casado, C. Electrónica y automática de bajo coste aplicada al huerto urbano. In Proceedings of the Congreso Ibérico de Agroingeniería, Huesca, Spain, 3–6 September 2019; Servicio de Publicaciones Universidad: Zaragoza, Spain, 2019; pp. 940–952.

22. Hormigo, J.; Rodriguez, A. Designing a Project for Learning Industry 4.0 by Applying IoT to Urban Garden. *IEEE Rev. Iberoam. Tecnol. del Aprendiz.* **2019**, *14*, 58–65. [CrossRef]

23. Khandakar, A.; Chowdhury, M.E.H.; Gonzales, A.J.S.P.; Touati, F.; Emadi, N.A.; Ayari, M.A. Case Study to Analyze the Impact of Multi-Course Project-Based Learning Approach on Education for Sustainable Development. *Sustainability* **2020**, *12*, 480. [CrossRef]

24. Derler, H.; Berner, S.; Grach, D.; Posch, A.; Seebacher, U. Project-Based Learning in a Transinstitutional Research Setting: Case Study on the Development of Sustainable Food Products. *Sustainability* **2019**, *12*, 233. [CrossRef]

25. Mantawy, I.M.; Rusch, C.; Ghimire, S.; Lantz, L.; Dhamala, H.; Shrestha, B.; Lampert, A.; Khadka, M.; Bista, A.; Soni, R.; et al. Bridging the Gap between Academia and Practice: Project-Based Class for Prestressed Concrete Applications. *Educ. Sci.* **2019**, *9*, 176. [CrossRef]

26. Blumenfeld, P.; Soloway, E.; Marx, R.; Krajcik, J.; Guzdial, M.; Palincsar, A. Motivating Project-Based Learning: Sustaining the Doing, Supporting the Learning. *Educ. Psychol.* **1991**, *26*, 369–398.

27. Fruchter, R. Dimensions of Teamwork Education. *Int. J. Eng. Educ.* **2001**, *17*, 426–430.

28. Thomas, J.W. A review of research on project-based learning. *Learning* **2000**, 1–46.

29. Cannon-Bowers, J.A.; Tannenbaum, S.I.; Salas, E.; Volpe, C.E. Defining competencies and establishing team training requirements. In *Team Effectiveness and Decision Making in Organizations*; Guzzo, R., Salas, E., Eds.; Jossey Bass: San Francisco, CA, USA, 1995; pp. 333–380. ISBN 9781555426415.

30. Hackman, J.R. A normative model of work team effectiveness. *Yale Sch. Organ. Manag.* **1983**.

31. Klimoski, R.J.; Jones, R.G. Staffing for effective group decision making: Key issues in matching people and teams. In *Team Effectiveness and Decision Making in Organizations*; Wiley & Sons Ltd.: Toronto, ON, Canada, 1995; ISBN 978-1-55542-641-5.
32. McGrath, J.E. Social Psychology: A Brief Introduction. *Sociol. Q.* **1964**.
33. Salas, E.; Dickinson, T.L.; Converse, S.A.; Tannenbaum, S.I. Toward an understanding of team performance and training. *Teams: Their Training and Performance* **1992**.
34. Tannenbaum, S.I.; Beard, R.L.; Salas, E. Team Building and its Influence on Team Effectiveness: An Examination of Conceptual and Empirical Developments. *Adv. Psychol.* **1992**, *82*, 117–153.
35. McGrath, J.E. *Groups: Interaction and Performance*; Cliffs, E., Ed.; Prentice Hall PTR: Upper Saddle River, NJ, USA, 1984; Volume 14.
36. Bravo, R.; Catalán, S.; Pina, J.M. Analysing teamwork in higher education: An empirical study on the antecedents and consequences of team cohesiveness. *Stud. High. Educ.* **2018**, *44*, 1153–1165. [CrossRef]
37. Pfaff, E.; Huddleston, P. Does It Matter if I Hate Teamwork? What Impacts Student Attitudes toward Teamwork. *J. Mark. Educ.* **2003**, *25*, 37–45. [CrossRef]
38. Gargallo, B.; Suárez-Rodríguez, J.M.; Pérez-Pérez, C. El cuestionario ceveapeu. un instrumento para la evaluación de las estrategias de aprendizaje de los estudiantes universitarios. *Reli. Rev. Electron. Investig. y Eval. Educ.* **2009**, *15*, 6. [CrossRef]
39. Kyndt, E.; Dochy, F.; Struyven, K.; Cascallar, E. The perception of workload and task complexity and its influence on students' approaches to learning: A study in higher education. *Eur. J. Psychol. Educ.* **2011**, *26*, 393–415. [CrossRef]
40. Sargent, L.D.; Sue-Chan, C. Does diversity affect group efficacy? The intervening role of cohesion and task interdependence. *Small Gr. Res.* **2001**, *32*, 426–450. [CrossRef]
41. Fransen, J.; Kirschner, P.A.; Erkens, G. Mediating team effectiveness in the context of collaborative learning: The importance of team and task awareness. *Comput. Human Behav.* **2011**, *27*, 1103–1113. [CrossRef]
42. Alavi, M. Computer-Mediated Learning: An Empirical Evaluation. *MIS Q.* **1994**. [CrossRef]
43. Lau, P.; Kwong, T.; Chong, K.; Wong, E. Developing students' teamwork skills in a cooperative learning project. *Int. J. Lesson Learn. Stud.* **2013**. [CrossRef]
44. Wageman, R.; Hackman, J.R.; Lehman, E. Team diagnostic survey: Development of an instrument. *J. Appl. Behav. Sci.* **2005**, *41*, 373–398. [CrossRef]
45. Nunnally, J.C.; Bernstein, I. *Psychometric Theory*; McGraw-Hill: New York, NY, USA, 1967; Volume 226.
46. Tabuenca, B.; Wu, L.; Tovar, E. The PRISMA: A Visual Feedback Display for Learning Scenarios. In Proceedings of the 13th International Conference on Ubiquitous Computing & Ambient Intelligence, Toledo, Spain, 5 December 2019; Volume 31, p. 81.
47. Naik, N. Choice of effective messaging protocols for IoT systems: MQTT, CoAP, AMQP and HTTP. In Proceedings of the 2017 IEEE International Systems Engineering Symposium (ISSE), Vienna, Austria, 11–13 October 2017; IEEE: Piscataway, NJ, USA, 2017; pp. 1–7. [CrossRef]
48. Kodali, R.K.; Sarjerao, B.S. A low cost smart irrigation system using MQTT protocol. In Proceedings of the 2017 IEEE Region 10 Symposium (TENSYMP), Cochin, India, 14–16 July 2017; IEEE: Piscataway, NJ, USA, 2017; pp. 1–5.
49. Tabuenca, B.; Sánchez-Peña, J.J.; Cuetos-Revuelta, M.J. El smartphone desde la perspectiva docente: ¿una herramienta de tutorización o un catalizador de ciberacoso? *Rev. Educ. a Distancia* **2019**. [CrossRef]

Article

A Scalable Architecture for the Dynamic Deployment of Multimodal Learning Analytics Applications in Smart Classrooms

Alberto Huertas Celdrán [1], José A. Ruipérez-Valiente [2,*], Félix J. García Clemente [2], María Jesús Rodríguez-Triana [3], Shashi Kant Shankar [3] and Gregorio Martínez Pérez [2]

[1] Telecommunication Software & Systems Group, Waterford Institute of Technology, X91 P20H Waterford, Ireland; ahuertas@tssg.org

[2] Faculty of Computer Science, University of Murcia, 30100 Murcia, Spain; fgarcia@um.es (F.J.G.C.); gregorio@um.es (G.M.P.)

[3] School of Digital Technologies, Tallinn University, 10120 Tallinn, Estonia; mjrt@tlu.ee (M.J.R.-T.); shashik@tlu.ee (S.K.S.)

* Correspondence: jruiperez@um.es

Received: 22 April 2020; Accepted: 19 May 2020; Published: 21 May 2020

Abstract: The smart classrooms of the future will use different software, devices and wearables as an integral part of the learning process. These educational applications generate a large amount of data from different sources. The area of Multimodal Learning Analytics (MMLA) explores the affordances of processing these heterogeneous data to understand and improve both learning and the context where it occurs. However, a review of different MMLA studies highlighted that ad-hoc and rigid architectures cannot be scaled up to real contexts. In this work, we propose a novel MMLA architecture that builds on software-defined networks and network function virtualization principles. We exemplify how this architecture can solve some of the detected challenges to deploy, dismantle and reconfigure the MMLA applications in a scalable way. Additionally, through some experiments, we demonstrate the feasibility and performance of our architecture when different classroom devices are reconfigured with diverse learning tools. These findings and the proposed architecture can be useful for other researchers in the area of MMLA and educational technologies envisioning the future of smart classrooms. Future work should aim to deploy this architecture in real educational scenarios with MMLA applications.

Keywords: smart classrooms; educational technology; multimodal learning analytics; internet of things; multisensorial networks

---

1. Introduction

Technology has been transforming education for the last decade. One of the main changes is the introduction of digital tools that support the learning and teaching practices [1]. Both software (e.g., smart tutoring systems, learning management systems, educational games, simulations, or virtual/augmented reality environments) and hardware (e.g., smart whiteboards, smartphones, remote labs, robots, wearable devices, cameras and other sensors) are present in the classroom and in our daily life [2,3]. The dynamism of classrooms requires the orchestration of this complex technical ecosystem, currently performed manually by instructors. Consequentially, novel technologies and mechanisms should be considered during the deployment of flexible and dynamic smart classrooms.

These rich ecosystems collect large amounts of data about the learning process and context, opening the door to better understand and improve education. However, handling such volume of raw data also represents a complicated challenge [4]. Aware of the promises and challenges, the area of Learning Analytics (LA) focuses on the "measurement, collection, analysis and reporting of data about learners and their contexts, for purposes of understanding and optimizing learning and the environments in which it occurs" (SoLAR definition of Learning Analytics https://www.solaresearch. org/about/what-is-learning-analytics). Within LA, over the last years, there has been a growing context on Multimodal Learning Analytics (MMLA), which is a sub-field that makes special emphasis on the usage of multimodal data sources [5]. There has been multiple and diverse MMLA applications, such as to teach how to dance salsa [6] or to assess oral presentations [7]. While transforming raw data into meaningful indicators is already daring [4], in this manuscript, we are mostly concerned with the issue of orchestrating the different data sources and applications. A recent literature review on MMLA architectures reveals that, due to the complexity of orchestrating the different elements of the technical ecosystem, most of the proposals offer ad-hoc solutions [8]. Apart from limiting the chances of reusability in different educational contexts, the effort to develop, deploy, maintain and enable interoperability among all those ad-hoc solutions does not scale up when the number of solutions increases [9]. Therefore, the current ad-hoc setup represents an important challenge to systematically apply MMLA in smart classrooms [10].

Thus, a real futuristic scenario with smart classrooms, where consecutive lessons take place (with 15–30 min breaks), would require a seamless and scalable reconfiguration of the sensors, devices and virtual learning environments within the classroom not only to deliver the lesson but also to profit from highly different MMLA solutions [10]. To address these challenges, we propose to evolve from traditional management, predefined by the instructor in a manual fashion, towards an automated approach able to reconfigure the classroom devices without human intervention and in a flexible and on-demand way. The number of sensors and actuators making up smart classrooms, as well as the possibility of managing them in a dynamic way make the scalability of the proposed approach a critical aspect to take into account. This can be possible by deploying a Mobile Edge Computing (MEC) architecture that combines Network Function Virtualization (NFV) technique [11] and Software-Defined Networking (SDN) paradigm [12]. NFV will allow for separating the software logic from the hardware of the classroom devices. It improves the flexibility and dynamism of device management processes by enabling the deployment, dismantling and reconfiguration of the technical ecosystem according to the current classroom needs. The SDN paradigm will help smart classrooms with automatic and dynamic management of network communications, enabling the Quality-of-Service (QoS) and interoperability of smart classroom devices and applications at the edge.

The objective of this paper is to present an MEC-enabled architecture that integrates SDN/NFV to deploy, configure and control the lifecycle of MMLA applications and devices making up a smart classroom as well as its network communications at any time and on-demand. More specifically, the objectives of this paper are as follows:

1. Use the MMLA literature to present a simulated but realistic scenario that can surface the limitations of the current technical approaches involved in the orchestration of complex technical ecosystems in educational practices.
2. Propose an MMLA architecture implementing SDN/NFV principles and exemplify how this architecture can solve some of the detected challenges to deploy, dismantle and reconfigure the MMLA applications in a scalable way.
3. Perform several experiments to demonstrate the feasibility and performance of the proposed architecture in terms of time required to deploy and reconfigure these applications.

The remainder of this paper is structured according to the next schema. Section 2 reviews and analyzes the state of the art of smart learning and classrooms, MMLA, remote smart classrooms, as well as the usage of SDN and NFV in different scenarios. Section 3 shows a case study explaining three different scenarios and their concerns. Section 4 describes the proposed architecture and how it can address the concerns of the aforementioned scenarios. Section 5 presents some experimental results that demonstrate the usefulness and performance of our solution. Section 6 discusses the main benefits of our solution compared to the existing ones. Finally, conclusions and future work are drawn in Section 7.

2. Related Work

2.1. Smart Learning Environments and Classrooms

In the last few decades, multiple terms have been coined with the "smart" label, often referring to devices (such as phones or watches) or spaces (e.g., classrooms, schools, campus, or cities) that through the utilization of the appropriate technologies and Internet of Things (IoT) services collect data from the users and the context to better adapt to the needs of the stakeholders involved. Aligned with this general idea, Smart Learning Environments (SLEs) are technology-enhanced learning environments able to offer instant and adaptive support to learners based on the analyses of their individual needs and based on the contexts in which they are situated [13]. Thus, when we think of a *smart classroom*, we should not reduce it to the mere idea of a traditional classroom heavily equipped with virtual learning environments and mobile, wearable or IoT devices.

While many aspects should be taken into consideration in a smart classroom such as the architectural design and its ergonomy, or the pedagogical methodology [14], in this paper, we focus on the infrastructure required to enable the "smart" features, i.e.,: (1) to seamlessly reconfigure such a complex technological infrastructure for guaranteeing the dynamicity and QoS of smart classrooms; and (2) to collect data from users and context to feed data for the intelligent adaptation to the learning needs at enactment time.

2.2. Architectures for Smart Learning Environments and Classrooms

As a recent literature review on smart campus technologies shows [15], paradigms and technologies such as the IoT, virtualization, wireless network, or mobile terminals are essential parts to be considered. There have been several attempts to orchestrate this intricate technical ecosystem. At the beginning, many of them were ad-hoc architectures suitable for specific technologies (e.g., interactive boards [16]), or focused on concrete problems (e.g., communication issues [17,18]) or features (e.g., remote software control [19]). Lately, authors have started broadening the scope and flexibility of their proposals. For example, GLUEPS-AR [20,21] combines the lessons learnt from distributed learning environments and the ideas coming from the MMLA domain. In [21], Serrano et al. designed an architecture which gathers student actions and their contextual data during across-spaces learning tasks to feed the adaption features. Another example is the architecture proposed by Huang et al. [22], which not only conducts the collection, integration and analyses of contextual data, but also enables the remote control of IoT devices and enhances the usability of the smart classroom with additional services such as voice recognition and user control interfaces. Previous colleagues also introduced LEARNSense framework [23], which aims to provide learning analytics using wearable devices. However, they did not deal with scalability and deployment issues either.

These architectures often focus on supporting data processing activities of the Data Value Chain (DVC) [24] (namely, collection and annotation, preparation, organization, integration, analysis, visualization, and decision-making). Each of these data processing activities poses a number of challenges linked to the problems associated with the data collection and analysis of multimodal data sources [8], which are common in smart classrooms. However, none of these proposals details how to (re)configure the smart classroom technical ecosystem to seamlessly switch from one LA application to another. Thus, in this paper, we try not only to enable the DVC in a smart classroom but also to reconfigure the technical ecosystem to cope with the requirements of different lessons happening in a row.

2.3. Remote Classrooms and Labs

Related to the technical orchestration challenges of smart classrooms, the virtual and remote lab field has a long trajectory coordinating IoT services and devices. Remote smart classrooms consider virtualization techniques and virtual machines (VM) to optimize the management of their software and hardware resources flexibly. Some remote laboratories consider virtual labs as an essential tool to improve the learning experience by supporting experimentation about unobserved phenomena [25]. In [26], the WebLab-Deusto project [27] used VMs to provide their students with remote smart laboratories that do not consider WebLab-specific code. Students had access to VMs for a given time and, once finished, a snapshot was made before restoring and preparing the VMs for new students. In [28], the authors proposed a solution that considered virtualization techniques to adapt the resources of remote laboratories at anytime and on-demand. Several experiments demonstrated how the usage of computing resources was optimized to guarantee the smart labs quality of service. In [29], the authors presented a mechanism to automatically generate, deploy and publish digitized labs in a framework of Massively Scalable Online Laboratories (MSOL). The authors demonstrated the suitability of the proposed mechanism by developing a communication protocol managing the lab equipment remotely, together with a web platform enabling the management of files and publishing digitized labs as web applications. Finally, the Smart Device Specification [30,31] provided remote labs with interesting capabilities. This specification focused on removing dependencies between clients and servers while enabling the description of remote lab experiments, and the selection of particular remote lab configurations [32]. However, configurations were not flexible enough because these must be established in advance by the lab administrator.

2.4. SDN and NFV Applied to Different Scenarios

The combination of SDN/NFV enables flexible, dynamic and on-demand management of networking and infrastructure resources. Moreover, it facilitates disruptive and heterogeneous scenarios such as the next generation of mobile networks (5G) [33], healthcare environments [34], or IoT [35].

Regarding 5G mobile networks, the authors of [36] analyzed the impact of SDN/NFV in the new vision of current and future network architectures. The authors highlighted how the combination of SDN/NFV reduces costs while improving the network flexibility and scalability of the infrastructure. The authors of [37] proposed a 5G architecture using NFV to support the implementation of tactile internet. A utility optimization algorithm which enables human perception-based tactile internet was developed to optimize the utility of 5G NFV-based components in this new scenario. In [33], the authors proposed an architecture which integrates SDN/NFV to manage and orchestrate services in charge of monitoring and controlling the network plane of a 5G network infrastructure in real-time and on-demand. Another solution was presented in [38], where authors studied the network flows migration of 5G networks and pointed out the inverse relationship between network load balancing and reconfiguration costs. Several experiments demonstrated the previous trade-off and the usefulness of the proposed solution. Regarding healthcare scenarios, the authors of [34] proposed an SDN/NFV architecture providing flexible and cost-efficient deployment and control of healthcare applications and services. In addition, the authors of [39] proposed an SDN/NFV framework to control the life-cycle and behaviour of physical and virtual medical devices belonging to clinical environments. This work also presented the novel concept of virtual medical device, an NFV-aware system providing dynamism in clinical environments. In the IoT context, the authors of [35] introduced an SDN/NFV architecture providing IoT devices with ultra-low communication latency. Another work was proposed in [40], where authors designed an architecture to ensure key security and privacy aspects of cyber-physical systems and IoT environments. SDN and NFV were considered to allow IoT devices and environments to make security decisions and take dynamic reactions. It is important to mention that learning scenarios such the proposed in this work can be improved by considering the SDN/NFV capabilities presented in the previous works.

In conclusion, this section has reviewed some of the most relevant solutions of heterogeneous smart learning environments and remote classrooms, highlighting the importance of seamless reconfiguration of smart classroom devices. The lack of solutions able to deploy, dismantle and reconfigure the software of classroom devices has also been demonstrated to ensure the seamless reconfiguration of devices in real-time and on-demand. Finally, we have shown the potential of SDN and NFV in other scenarios to achieve flexible and dynamic management of computational, storage and networking resources.

3. Description of Simulated Case Study

The related work review concluded that one of the main challenges in the area of smart classrooms and MMLA context is an architectural one. In our attempt to understand in depth this research issue, this section presents a simulated case study inspired by authentic uses cases extracted from the literature. The main goal is to ascertain what the specific issues are that our proposed architecture must address (see in next Section 4), in order to support a seamless reconfiguration of a smart classroom where different learning activities would happen in a row. With the objective of building this case study, we reviewed literature on MMLA applications that have been implemented during the last few years. From these cases, we select three that were aligned with innovative learning trends and have different objectives, devices, analytics and sensors in order to demonstrate how the architecture self-organizes from one scenario to the following one. We also order these three cases by increasing complexity, the first one focuses on individual students, the second one focuses on groups of students collaborating, and the third one focuses on students collaborating in projects but also on what the instructor is doing. Next, we describe in depth each one of the scenarios.

3.1. Intelligent Tutoring System in the Classroom

One of the main trends in education over the last decade has been the development of interactive environments that can be slowly introduced as part of the classroom or homework activities. Two of the most relevant tools for this purpose are Intelligent Tutoring Systems and Educational Games [41]. Most of the literature meta-reviews that have measured the effectiveness of such tools in the classroom [41,42] have reported positive effects. However, these studies also agree on the struggle that instructors face to effectively integrate these tools in their teaching and curriculum. One of the reasons is not being able to know what students are doing in these virtual environments to orchestrate the classroom activities and to intervene if necessary. Hence, the need for the development of real-time dashboards that can provide this information to instructors [43].

The first scenario is grounded in this technological and pedagogical issue, and is strongly inspired in the previous work of Holstein et al. [44,45]. In this work, they have co-designed a dashboard and augmented wearable instruments to visualize real-time analytics and visualizations of what each student is doing in the intelligent tutoring system. Next, we detail the specific details:

- **Context**: In this scenario, students are practicing a specific topic through the use of an intelligent tutoring system. Each student is individually interacting with the environment with the computer. In order to provide just-in-time help, instructors need to know how students are advancing in this practice and what are their mistakes or misconceptions. A usual class would be around 20 to 40 students.

- **Application**: When students interact with the intelligent tutoring environment, they generate events and clickstream data that can be processed to make inferences about their learning process. Based on these data, the analytics engine generates a number of indicators of students' current skill and behavioral states. For example, it can show if a student is confused, needs help, has been idle for a number of minutes or their areas of struggle, among other pieces of information. Additionally, each computer has a webcam capturing students' face and expression, and the analytics engine applies an affect detection Machine Learning (ML) model to infer students' affect status. Instructors receive all these info through a dashboard in real-time and can easily move within the classroom attending students' needs.

- **Sensors and devices**:

- *Individual students' devices*: Students interact with the ITS by connecting to it as a web application. The ITS provides of a series of scaffolded exercises adapted to the current level of skill of each student. Students use the desktop PC available in the classroom.
- *Individual students' webcam*: Each student has a front camera in their computer that is capturing a video feed of their face expression continuously. This feed is used by the analytics engine to infer the emotional state in time windows.
- *Instructor device*: The instructor consume the analytics via a dashboard by connecting from its device (tablet or laptop) to the visualizer provided by the architecture.

3.2. Tabletop Task Collaboration

UNESCO has noted that the future of education should be focused on promoting transverse skills, such as collaboration [46]. The trend has shifted from individual efforts to group work, making the development of collaboration skills mandatory with an increasing trend of implementing collaborative learning activities with high frequency [47]. Therefore, it is not a surprise that numerous researchers have started to analyze collaborative learning from different perspectives. However, one of the challenges has been to scale up the analysis of these collaboration studies when there are many groups to assess or to provide feedback in real-time. Hence, the area of MMLA has been studying ways to automatically provide empirical evidence that can help to support co-located collaboration through analytics [48]. In these studies, researchers capture multimodal data from the collaboration, some examples of data sources include video, audio, physiological signals using wearables or interaction data with computers or shared devices [49,50].

This second scenario is grounded in this context where we present an application that generates colocated collaboration analytics while students are interacting on a multi-touch tabletop doing a collaborative task, which is based on previous work from Maldonado et al. [51]. The details of this scenario are depicted next:

- **Context**: In this case scenario, we have students interacting with a shared device known as interactive multi-touch tabletop, which can easily support face-to-face collaboration with multiple students interacting at the same time. Students carry out an activity on collaborative concept making, which is a technique where learners represent their understanding about a topic in a graphical manner by linking concepts and preposition [52]. At the same time, students are also conversing with each other and discussing their decisions, and this voice stream is also captured through a microphone. The class is organized in groups of three students, and a usual class could have around 7 to 14 groups.
- **Application**: The objective is to design an application that can help teachers become more aware of the collaborative process, by making visible interactions that would otherwise be hard to quantify or notice. The application study collaboration by considering both the verbal interactions when students are talking to each other, as well as physical touches with the table-top [53]. More specifically, it can use metrics to identify learners that are not contributing enough to the activity or are dominating it (both physical and verbal interaction), groups that can work independently or those that do not understand the task. The instructor accesses all these information though a visualization dashboard in a hand-held device.
- **Sensors and devices**:

  - *Group multi table-top*: Table-top learning environments are big tactile screens that allow the collaboration of multiple users at the same time.
  - *Group overhead depth sensor*: A Kinect sensor is used to track the position of each user automatically detecting which student did each touch.
  - *Group microphone array*: It is located above the tabletop and captures the voice of all the group members, distinguishing the person which is speaking.

- *Instructor device*: The instructor consume the analytics via a dashboard by connecting from its device (tablet or laptop) to the visualizer provided by the architecture.

3.3. Programming Project-Based Learning and Instructor Indoor Positioning

Project-based learning has become one of the main forms of instructions across contexts and the different phases of schooling as it resembles better real-world practices and leads to deeper learning [54]. This method of instruction is very common in programming courses, where students often have to develop a collaborative group programming project to pass the course (e.g., [55]). One of the challenges of these collaborative projects is to assess the role and effort of each member of the group in order to guarantee similar workload distribution, hence avoiding free riding [56]. These project-based learning courses often have entire sessions devoted to in-class work on the projects. During these sessions, the teacher moves from group to group solving doubts, which presents a new challenge regarding how to equitably distribute their time across groups [57]. In this context, we can collect diverse sources of data from the collaborative programming environments, audio from group conversations, instructors' position and physiological signals from the students.

This third scenario combines inspiration from the following previous studies: the work of Spikol et al., and Blikstein to apply MMLA to analyze collaborative project-based learning and open-ended programming tasks [58,59], the ideas of Ahonen et al., to analyze biosignals during these programming tasks [60] and finally the proposal of Martínez-Maldonado et al. to estimate the amount of time spent by the instructions in each group [57]. Therefore, in this scenario, the application combines an analysis of the collaborative programming actions and conversation of each group, the physiological signals levels of each student and position of the instructor. More details about this scenario are depicted next:

- **Context**: Numerous programming courses have capstone projects where students need to implement an application that shows evidence of the different concepts acquired thorough the course. These courses usually have some sessions allocated for students to start developing these projects in groups while instructors move from one group to another solving doubts. Each group interacts with a shared programming environment (e.g., [61]) to develop the project collaboratively. The class is organized in groups of three students, and a usual class could have around 7 to 14 groups.
- **Application**: In this scenario, there are two main applications. The first one is to provide analytics regarding how the collaboration is working out and how the project is advancing. This can include information regarding areas of struggle based on the code written and code compilations [59], but also regarding the level of contribution to the project of each member, analysis of the conversation and engagement levels obtained through the analysis of the physiological signals to measure activation and engagement levels. The second one is an automatic control of how much time the instructor has spent helping each one of the groups through indoor positioning; this way, the instructor can balance the help that each group receives. The instructor can consult all this information through a dashboard in order to provide just-in-time and personalized support to each group.
- **Sensors and devices**:

  - *Individual students' devices*: Students interact with the collaborative programmings environment by connecting to it through a web application.
  - *Individual Empatica E4 wristband*: Each student wears an E4 empatica wristband that captures the heart rate, a three-axis activity through an accelerometer, and the electrodermal activity of their skin.
  - *Group microphone array*: It is located above each one of the groups' tables, distinguishing the person which is speaking.

- *Group positioning sensor*: It is located in each one of the groups' tables to detect the center position of each group.
- *Instructor positioning badge*: It is carried by the instructor when moving around the class. It implements Pozyx (https://www.pozyx.io/) technology which is an ultra wide band solution that provides accurate positioning and motion information with sub-meter accuracy (10 cm).
- *Instructor device*: The instructor consumes the analytics via a dashboard by connecting from its device (tablet or laptop) to the visualizer provided by the architecture.

3.4. Requirements of the Previous Scenarios

The case study with the three consecutive scenarios represents an example of how smart classrooms and MMLA solutions could look in the future. To reach our goal of supporting the seamless reconfiguration and data collection required to enable the smart adaptation, we have identified four main requirements emerging from our simulated case study:

Requirement 1—Within-scenario flexibility for instructor-configured data collection, analytics, visualizations, and recommendations: Aligned with the challenges reported in the literature [8], the MMLA solutions implemented in the aforementioned scenarios are ad-hoc solutions that enable the data gathering and analysis to later feed the visualizations and recommendations for instructors and students. The three use cases that we described have different learning environments, devices, data sources or analytics pipelines that have been configured to match the necessities of each use case. Therefore, to be able to scale up the number of MMLA solutions used in a single classroom and scenario, it is necessary to provide a scalable architecture compatible with the different MMLA applications [9,10] by abstracting these functionalities in scalable and interoperable modules that can be automatically re-configured for each MMLA application.

Requirement 2—Between-scenario flexibility for automatic deployment of the MMLA solutions: The kind of equipment, devices, setup and sensors necessary to perform these applications makes smart classrooms expensive to have. Therefore, we would expect that, in the future, these classrooms are fully booked, perhaps having a short time of 15–30 min in-between sessions. In our case study, we presented three consecutive use cases to illustrate this issue, but this might be a conservative estimate. The current setup makes it very challenging to seamlessly and automatically re-configure the technical ecosystem and to also enable the data collection and analysis in short periods of time. In our case study without a proper architecture, each teacher would be in charge to deal with the technological complexity of the MMLA application in each class, which in reality is not a feasible approach. This raises the necessity to have a seamless transitions between the scenarios of our simulated case study.

Requirement 3—Seamless privacy and authentication configurations: The privacy of users, and of students in this case scenario, has been one of the topics on the spotlight during the last years [62]. The regulations have agreed that we need to provide control to the users so that they can specify how their data can be used. Therefore, even though these MMLA solutions seek to help students in their learning process, students and instructors should still have the right to opt-in or -out so that their data are not collected and/or used. In the case scenario, each application would need to manage this privacy and authentication issues separately, which is sub-optimal. Therefore, we need to provide a centralized system where students can configure their privacy and authentication options to apply across all the smart classroom applications, and we also need to easily identify students across applications and devices so that we can properly process their data.

Requirement 4—Easy communication with external data sources: Thanks to the institutional data and the ICT adoption in our daily routines, there can be numerous data sources (both formal and informal) that can hold valuable information to understand students' context and knowledge. Some examples might include the classical LMSs in formal learning institutions, other online courses, academic records or background information. In the case study, each application would have developed their own interface to interact with these external data sources. Thus, instead of implementing ad-hoc solutions to benefit from those external data sources, there is a need for generating services and APIs that can be used across applications.

4. Architecture

This section describes our MEC/SDN-oriented architecture that satisfies the aforementioned requirements, and how it integrates different components to reconfigure and manage the learning applications running on top of classroom devices automatically, on-demand and in real-time. Figure 1 shows the levels, components and communications of the proposed architecture. The main elements, following a top-down approach, are the next ones:

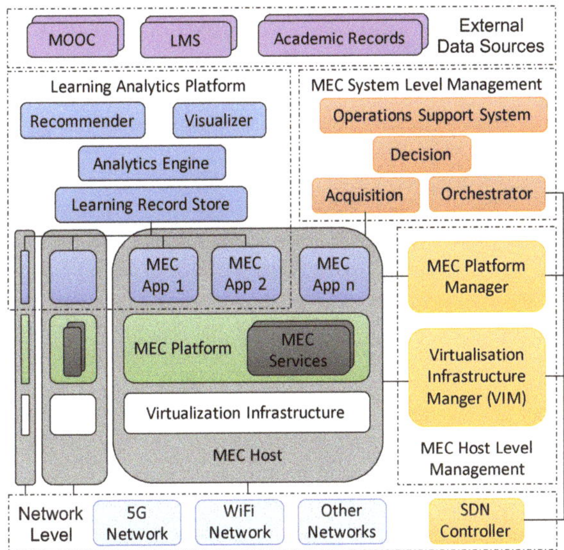

Figure 1. Architecture oriented to the Mobile Edge Computing (MEC) paradigm.

- *External Data Sources.* This level contains different external databases and tools such as data from the Academic Records, Learning Management System (LMS) or Massive Open Online Courses (MOOC) that can feed our architecture with relevant students' data.
- *Learning Analytics Platform.* It hosts the components focused on analysing data provided by external sources and generated during the realization of learning activities.
- *MEC System Level Management.* This level is focused on (1) processing requests from instructors to reconfigure heterogeneous classroom devices in real-time and on-demand, (2) making decision and orchestrating them to configure learning applications running on top of classroom devices, and (3) sensing classroom devices to detect misconfigurations or problems.
- *MEC Host.* Heterogeneous classroom devices, also known as MEC Hosts, such as electronic blackboards, tablets, personal computers, servers, or Raspberry Pi that need to be reconfigured according to the current learning course or subject.
- *MEC Host Level Management.* Level hosting the different managers able to control the life-cycle of the Virtualization infrastructure, MEC Platform, and MEC Apps running on the MEC Hosts.
- *Network Level.* This level contains the network infrastructure enabling the communication of MEC Hosts and the rest of the levels making up the architecture.

In the following subsections, we explain in detail the components and main levels of our platform.

4.1. Learning Analytics Platform

The *Learning Analytics Platform* has the different modules and components that are necessary to implement learning analytics applications that have as a final objective to improve the learning experience and outcomes of students. With that goal in mind, the platform hosts different components able to acquire, process, analyze, recommend and visualize relevant data generated during the interaction of students with learning applications. Among the most relevant components, we highlight the *Learning Record Store (LRS)*, which acquires and stores students' interaction registers generated by learning applications. Those registers are sent to the *Analytics Engine* component to analyze them by using ML and statistical techniques. According to the registers, the outputs of the *Analytics Engine* and some trained models, the *Recommender* component provides students and instructors with suggestions to improve the learning experience. Finally, the *Visualizer* component exposes a graphical interface that allows students and instructors to interact with registers, data and outputs of the learning platform.

4.2. MEC System Level Management

The *MEC System Level Management* deals with the management of the classroom devices and the behaviour of the learning applications running on top. In this context, the *Operation Support System* (OSS) is focused on the the logic of the architecture. This element provides instructors with an interface to define the rules that enable the reconfiguration of the learning applications and software running on top of the heterogeneous devices belonging to a classroom. These rules will be provided to the *Decision* component to identify particular actions to be taken. Once a decision is made, the *Orchestrator* receives the notification and interacts with the managers and controllers of the lower levels to configure the network, the classroom devices and their learning applications. Finally, the *Acquisition* component senses data generated by the classroom devices and their applications and services (not only learning applications) to detect misconfigurations or problems. When one problem is detected, the *Decision* and *Orchestrator* modules come into play to decide, schedule, and spread the required actions.

4.3. MEC Host Level

The *MEC Host Level* is composed of two planes, the control and data planes.

The control plane is called *MEC Host Level Management* and it is in charge of deploying, controlling and dismantling learning applications, instantiated as *MEC Apps* that run on top of heterogeneous classroom devices (*MEC Hosts*). The MEC Host level management contains two managers: the *MEC Platform Manager* and the *Virtualization Infrastructure manager* (VIM). The MEC Platform Manager controls the whole life-cycle of MEC Apps, and the VIM manages the computation, storage and networking virtual and physical resources of the Virtualization Infrastructure.

In the data plane, we find the MEC Hosts, which are classroom devices providing computational, storage, and networking resources to execute learning applications. Each MEC Host contains a *Virtualization Infrastructure*, a *MEC Platform* and one or more *MEC Apps*. MEC Apps can be deployed as learning applications, components of the Learning Analytics Platform (commented on in Section 4.1) and other applications like, for example, those oriented to improve the learning courses security and privacy). MEC Apps can be instantiated in Virtual Machines (VM) or containers running on top of the virtual infrastructure. The virtualization infrastructure consumes the hardware of heterogeneous learning devices such as computers, digital blackboards, or cameras and provides computational, storage and networking virtual resources. Finally, the MEC platform provides essential and generic MEC Services needed to run MEC Apps. These services can be specific for particular applications or generic enough to be shared among several MEC Apss. Examples of MEC Services can range from communication protocols to access control mechanisms or cryptographic material.

4.4. Network Level

The *Network Level* contains two types of elements: heterogeneous *Networks* and the *Network manager*. The networks represent the hardware and software networking resources needed to connect MEC Hosts and their MEC Apps. The Network Manager allocates the SDN Controller, which has the global view of the network status as well as the logic of the network to control the data plane where heterogeneous networks are located.

4.5. Solutions Provided by our Architecture to the Previous Requirements

Solution to Requirement 1—Within-scenario flexibility for instructor-configured data collection, analytics, visualizations and recommendations: Easy and flexible reconfigurations of the instructors' and learners' applications, such as the one needed in the first scenario, are enabled by our solution. Figure 2 shows the interaction between the components of our architecture to reconfigure the storage and processing capabilities of the instructor host. For clarity's sake, we show how the architecture reconfigures two MEC Apps running on top of an MEC Host. However, this functionality could be extended to several MEC Hosts and applications. The 1st step of Figure 2 shows when the decision of reconfiguring the instructor host is made by the Decision component. After that, the Orchestrator provides the MEC Platform Manager with the MEC Host and the reconfiguration details of the new storage and processing capabilities. Once received, the MEC Platform Manager interacts with the instructor host to access the storage and processing MEC Apps and reconfigure them (steps from 3 to 6). When the reconfigurations have finished, the action is confirmed to the Orchestrator (step 7).

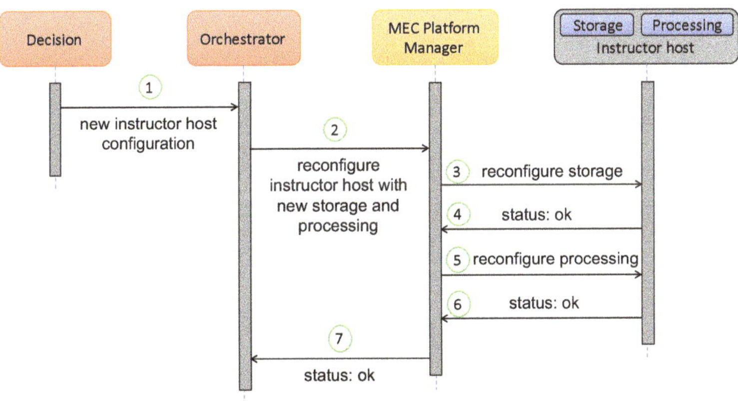

Figure 2. Architecture reconfiguring two MEC Apps running on top of an MEC Host.

Solution to Requirement 2—Between-scenario flexibility for automatic deployment of the MMLA solutions: Aligned with the capabilities shown in the previous issue and focused on addressing this one, the proposed architecture deploys, configures and dismantles MEC Hosts and their applications in real-time and on-demand. Following the previous example, Figure 3 shows how the components of our architecture dismantle the instructor host when a given application is finished, and deploy new ones with different capabilities for the next class. In the 1st step of Figure 3, the Decision component interacts with the Orchestrator to notify the necessity of changing the instructor host. After that, the Orchestrator provides the VI Manager with the required info to dismantle the MEC Host (step 2). Once the notification is received, the VI Manager dismantles the host and confirms the Orchestrator the action (steps 3–5). When the old instructor host has been dismantled, the next step is to deploy a new MEC instructor host with more hardware resources (processor and graphics). This process is shown from 6 to 9 in Figure 3. At this stage, our architecture has already deployed a new

MEC instructor host with enough hardware resources to meet the requirements of the next learning analytics application and the next step is to deploy a new MEC App with visualization tools and capabilities. For that, the MEC Platform Manager is the component in charge of deploying, configuring and confirming the new MEC App (steps from 10 to 13). Finally, the Orchestrator communicates with the SDN Controller to include a new rule in the switch flow table, and route the network packets sent and received by the new instructor host and its applications (step 14).

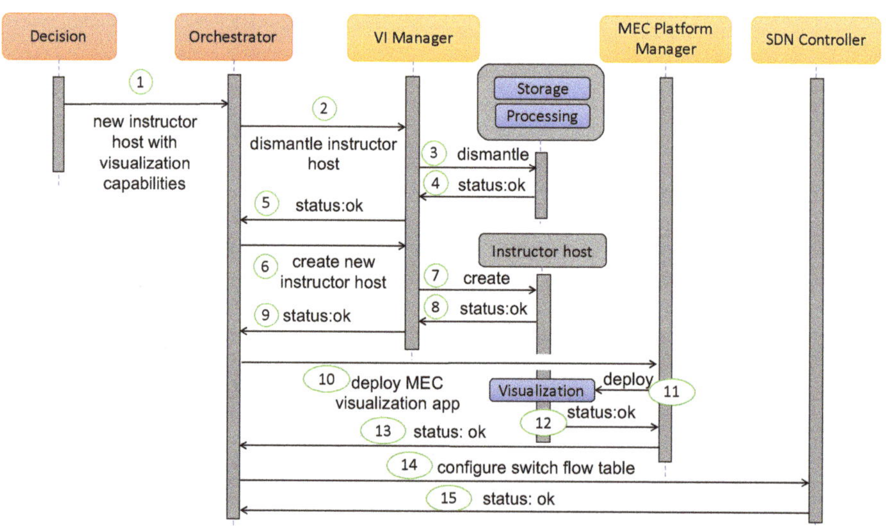

Figure 3. Architecture dismantling an old MEC Host, and deploying a new MEC Host and MEC App.

Solution to Requirement 3—Seamless privacy and authentication configurations: Our architecture is able to deploy MEC Apps, providing students with authentication and authorization capabilities, in real-time and on-demand. On the one hand, depending on the learning course security requirements, the architecture will deploy and configure an MEC App providing several authentication mechanisms with different levels of security. On the other hand, the architecture will deploy another MEC App allowing students to define their privacy preferences by defining user-friendly policies. In this context, students will determine what pieces of sensitive data can be shared, who or what learning tools can process the sensitive data, how long data can be processed or stored, or what can be done with the data, among others. Once defined the policies, they will be sent to the components of the Learning Analytics Platform to ensure that they are considered during the data management and storage processes.

Solution to Requirement 4—Easy communication with external data sources: As can be seen on top of Figure 1, the design of our architecture considers external data sources such as MOOC, LMS, or academic datasets feeding the Learning Analytics Platform with additional data that will be critical for the data analysis processes performed by its components.

5. Experimentation Results

A key aspect of our proposal is how the architecture deploys and configures the learning ecosystem automatically for each scenario, which addresses the aforementioned Requirements 1 and 2. We consider these two requirements as the key ones that are necessary to bring scalability and interoperability to smart classrooms and MMLA applications, and thus we focus our experimentation in this section on those two aspects. The deployment process is dealt by the Orchestrator that must consider the features of each classroom device and its performance with different MEC Apps. In this section, we show experimental results regarding computational performance and efficiency of typical classroom devices with practical learning tools.

With a model of deploying MEC Apps based on containers, we investigate experiments about three types of learning tools: high-intensive computing, medium-intensive computing and high-intensive data consuming. The high-intensive computing MEC App is a face-recognition that detects all the faces and face encoding in each frame of a video source. This application is a Python program based on *dlib* library using a Histogram of Oriented Gradients (HOG) face detector. The medium-intensive computing MEC App is a feature extractor for Automatic Speaker Recognition (ASR). This application is also a Python program based on the Mel-Frequency Cepstral Coefficients (MFCC) that analyzes an audio source periodically each second. In addition, the high-intensive data consuming MEC App is a computational physics simulation that plots a 3D surface. This application is a Python program based on Matplotlib library for creating animated visualizations.

5.1. Testing Environment

We deployed a testing environment composed of three MEC Hosts with different hardware resources, which are representative of a real smart classroom; these are a server, a desktop PC and a laptop. These devices can be used in the different scenarios presented in Section 3. The server was an Intel machine with dodeca-core (12 cores) 3.50 GHz CPU and 32 GB DDR4 RAM, the PC was an Intel machine with octa-core (8 cores) 3.40 GHz CPU and 16 GB DDR4 RAM, and the laptop was an Intel Celeron machine with dual-core 1.10 GHz CPU and 4 GB DDR4 RAM. In particular, laptops have similar computational capabilities to tablets and mini-PCs, so our experimental results with laptops are comparable to tablets and mini-PCs.

For each device, we set up a realistic evaluation environment with the typical services and graphical interface used to reduce the overhead. The operating system of all hosts was Ubuntu 64-bit 18.0.4, and the containers were deployed by the latest version (19.03.6) of Docker Engine. No more additional software components were needed to deploy the learning tools on our testing environment. Each learning tool was allocated within a unique Docker container providing a single learning task.

Our testbeds evaluated the performance and efficiency of our solution by increasing the number of containers on each type of MEC Host. This allows for observing the performance variance across different scenarios according to their capabilities. We expect that changing between scenarios would have an impact in the performance, e.g., the learning device installed in a classroom work table would require much more learning tools in a Tabletop Task Collaboration scenario than a Programming Project-based Learning scenario. Another possibility is that there could be changes in the number of students taking each class, hence affecting the computation requirements. Therefore, the performance for each configuration must be well-known by the Orchestrator to properly reconfigure the learning devices in each class.

5.2. Docker Container with High-Intensive Computing Application

There are several learning scenarios that can require a face detection tool to identify students or infer affect states. As shown in Section 3 for an Intelligent Tutoring System, an MEC Host with a camera capturing a video feed of student face expression can be used to infer the affect (e.g., surprise, neutral, confusion and angry) and identify when a student needs help. We used *dlib* library to implement a HOG face detection MEC App and created a Docker container that provides this app in our testing environment. The HOG is one of the most reliable and applied algorithms for person identification, but also an intensive computational task. Therefore, it is essential to properly manage the available computing resources in the learning device that can be dedicated to the execution of this learning tool.

In order to evaluate the performance and efficiency of the Face Recognition application in Docker containers, our testbed used a H.264 video source with 640×360 image size and applied the HOG algorithm in each video frame. We used the analyzed frames per second (FPS) as performance evaluation index to assess how fast the HOG algorithm is. If a configuration has higher FPS value, it has higher video quality and can produce smoother video. Figure 4 shows the experimental results obtained when increasing the number of containers for each type of learning device. The left graph depicts the maximum analyzed FPS for each configuration and the right graph shows how many CPU cores are overloaded.

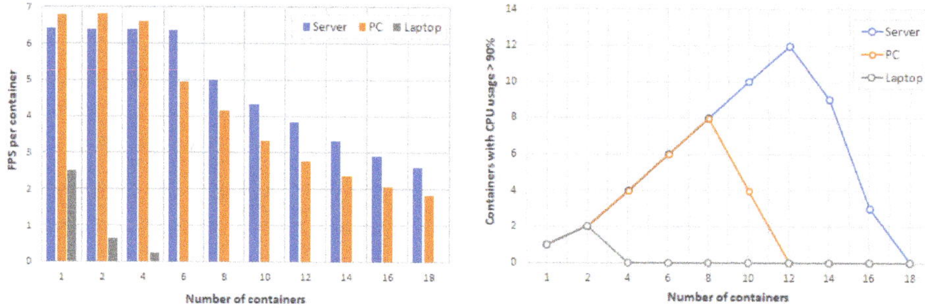

(a) Face detection and encoding speed.   (b) CPU usage per container.

Figure 4. Performance results for Face Recognition application in Docker containers.

As it can be seen in Figure 4, the maximum speed achieved was above 6 FPS for configurations with up to six containers in server and up to four containers in PC, whereas the throughput in the laptop was much lower with less than 3 FPS. In addition, the server was absolutely overloaded with 12 containers, the PC with eight containers and the laptop with two containers. Therefore, we observe that each container consumed approximately one CPU core. These experimental results imply that a face detection tool can be provided in different configurations e.g., a PC with eight cameras could serve for a work table shared by eight students or a laptop for one single student. Note that the server achieved the highest computation performance, and this performance could further improve if it included a graphics card to implement the HOG algorithm.

5.3. Docker Container with Medium Computing Application

Identifying students via their voice in a microphone can be useful for several learning scenarios, as shown in our use case related to project-based learning (see Section 3). An MEC Host with a microphone capturing the meeting audio can identify students, perform speech-to-text transcription, calculate speaker metrics (e.g., speaking time or counters) and infer the emotional state (e.g., angry, boring or excited).

We implemented an MEC App based on MFCC to recognize persons and created a Docker container with this tool to carry out our experiments. The MFCC are widely used in automatic speech and speaker recognition and allow transforming the audio source into a sequence of feature vectors that characterize voice signals. Our MEC App extracted feature vectors in one second window in order to apply a real-time student recognition. The process to calculate MFCCs consisted in framing the signal in short windows to later apply specific mathematical operations that convey a medium computing task.

In order to evaluate the performance and efficiency of the ASR application in Docker containers, our testbed used an audio signal stored and processed each second using MFCC with a length of the analysis window of 25 ms and a step between successive windows of 10 ms. In this case, we used the processing time as the performance evaluation index because this indicator shows how fast the MFCC algorithm is. If a configuration has slower processing time, it can process more audio sources and serve more users. Figure 5 shows the experimental results obtained when increasing the number of containers for each type of MEC Host. The left graph depicts the processing time to analyze the audio signal each second and the right graph shows how many CPU cores are used in the processing.

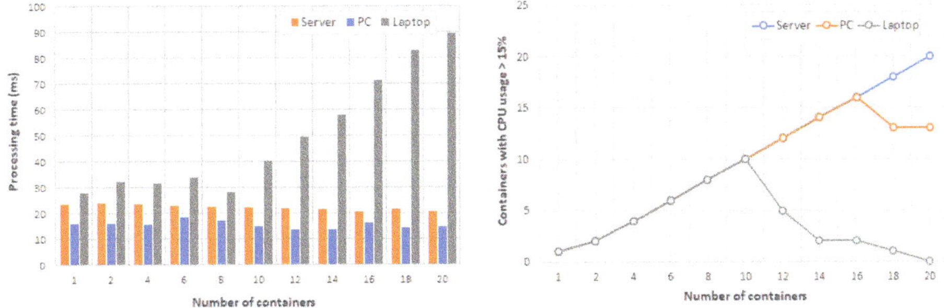

(a) Feature extractor speed.   (b) CPU usage per container.

Figure 5. Performance results for Automatic Speaker Recognition application in Docker containers.

The audio feature extractor is a relatively low computationally expensive task that is well-supported in server, PC and even laptop. As shown in Figure 5, the processing time was always below 100 ms for our three learning devices and below 30 ms for server and PC. However, the CPU overload was relevant for PC when the number of containers doubles its number of cores. In addition, the laptop was stuck when the number of containers was greater than 10, whereas the server was not overloaded with up to 20 containers. These experimental results show that an ASR tool can be easily provided in our use cases, e.g., a laptop/tablet with microphone could serve a 6-student work group or a server with 20 students simultaneously.

5.4. Docker Container with a High-Data Consuming Application

The interactive simulation-based learning can be useful in multiple scenarios, for example using an ITS in the classroom, as shown in Section 3. When students interact with the simulation, they generate events and clickstream data that can be stored and processed to calculate usage metrics (e.g., idle times or event counters) and even infer about their learning experience (e.g., difficulty or simplicity).

There are several types of interactive simulations which could be used in a classroom. Physics simulations are widely used to improve the learning process in science and engineering education. We implemented a Matplotlib MEC App to build an animated physics simulation that shows a wave motion. In particular, the physics simulation used a 1.5 GB array to plot a 3D surface animated. The size of plotting array implied that the simulation carried out a high data consuming task for learning devices or MEC Hosts.

In order to experiment our physics simulation in Docker containers, our testbed updated the plotting constantly in order to evaluate the performance in each learning device. We used the changes per second (CPS) of the simulation as the performance evaluation index because this indicator shows how fast the simulation is running. If a configuration has higher CPS value, it has higher simulation quality and can produce more fluent simulations. Figure 6 shows the experimental results obtained when increasing the number of containers for each type of learning device. The left graph depicts the maximum CPS for each configuration, and the right graph shows the percentage of RAM memory used.

Given the results shown in Figure 6 and that our physics simulation required 4 CPS at least to show a fluent animation, a laptop served only one container with our simulation. However, the server and PC could achieve 20 and 14 containers, respectively. Moreover, the memory was full and additional containers were rejected when the server launched more than 20 containers, the PC 14 containers, and the laptop 2 containers. These experimental results show that a high-data consuming simulation can be used in different configurations, e.g., a laptop/tablet could be used for a single student and a server to up to 20 students in the classroom.

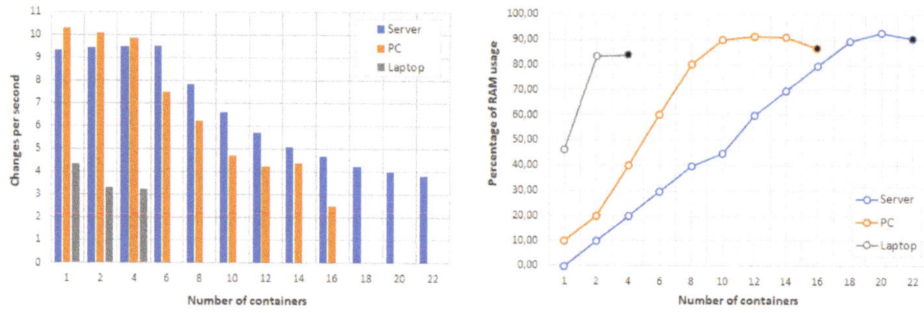

(**a**) Simulation speed.   (**b**) RAM memory usage per container.

Figure 6. Performance results for Computational Physics simulation in Docker containers.

6. Discussion

Among the different aspects to be taken into consideration in a smart classroom [14], the proposed architecture focuses on orchestrating the complex technical ecosystem and enabling its "smart" features. The architecture has been designed bearing the following main requirements in mind: within-scenario and between-scenario flexibility, seamless privacy and authentication configurations, and easy communication with external data sources.

The experimental results regarding the performance and scalability of our architecture show how heterogeneous classroom devices can be managed in an automatic and efficient way to host different amounts and types of learning tools and applications. Concretely, we demonstrated the scalability of our architecture when an increasing number of Dockers, with diverse computational requirements, is deployed over three widely used hardware configurations such as laptops, personal computers and servers. However, no direct comparison of the obtained results with those reported in the literature was possible since they highly depend on the hardware and software configuration. Furthermore, most MMLA studies evaluate their results based on educational outcomes but not on technical performance.

The automatic and flexible management of the proposed architecture has been motivated through the case study presented in this paper, which illustrates the limitations of current solutions and how our proposal offers a seamless switch between three different learning scenarios happening in the same smart classroom. While existing architectures for smart classrooms often involve ad-hoc digital devices and tools that can be used in specific ways [15,22], in our proposal, the different modules of the ecosystem can be orchestrated for multiple purposes in scalable and interoperable ways. Moreover, the human intervention required to adapt and reconfigure the transition between heterogeneous learning lessons is significantly reduced and can be automatized.

The presented architecture could be of great value also for the remote lab community. While virtualization techniques had been already explored [27–29], this architecture could increase the flexibility of remote labs, by supporting the configuration and deployment of remote experiments [32]. Moreover, it supports the collection of multimodal data (coming both from hardware and software) necessary to support the smart adaptation to the learning process.

Regarding the instant and adaptive support expected from smart classrooms [13], our proposal could become the base upon which other architectures could build, uncoupling the multimodal challenges of the DVC [4,63]. More concretely, our contribution helps to address the lower level technical requirements of the DVC, and more conceptual architectures (e.g., [21,22,63]) could build on top of it. Thus, our proposal contributes to diminishing the need for ad-hoc MMLA solutions often due to the technical constrains to the ecosystem [8]. As a consequence, relying on a lower level architecture will open the door to multiple analysis and adaptability schemes in smart classrooms, addressing the reusability and interoperability problems among MMLA solutions [9,10].

The integration of SDN/NFV in our architecture allows instructors to reduce their workload avoiding the manual configuration of classroom devices according to the topic and purpose of each subject. It also reduces the complexity of the smart classrooms management as well as optimizes the usage of classrooms devices. In a nutshell, smart classrooms equipped with our architecture will be able to reconfigure and optimize the learning applications of their devices ant their communications according to the current subject topic and number of students. It will be done in real-time and on-demand. In contrast, as it has been demonstrated in Section 2, existing solutions using virtualization techniques [25,26] are not able to reconfigure the whole remote lab in a flexible way. They just consider predefined VMs implementing particular learning applications that are instantiated and dismantled. It means that they miss critical aspects such as the flexible management of communications, essential to guarantee QoS issues when the number of students increases, and the optimization of hardware resources of learning devices such as CPU, memory and storage.

It is important to note that one of the main limitations of the proposed architecture is the complexity of its deployment. The usage of resource-constrained devices such as digital boards or cameras makes very complex their management through current virtualization techniques. Fortunately, this issue is limited when other devices such as tablets and personal computers are considered in smart classrooms. Additionally, the architecture is still to be tested in a real scenario, which is part of the future work. Moreover, we argue that the architecture represents an improvement with respect to other studies. However, we cannot present a direct comparison in terms of efficiency because most MMLA studies do not report on the performance of the architectures from the technical point of view. Finally, we still have not tackled the challenge of how instructors will be able to interact with this architecture through a user-friendly authoring tool.

7. Conclusions and Future Directions

Smart classrooms require a dynamic and flexible orchestration of their complex ecosystem, currently performed manually by instructors that use ad-hoc learning applications. With that goal in mind, this paper the following three key research problems: (1) the limitations of current learning solutions in terms of flexible and scalable management of devices belonging to simulated and realistic learning scenarios; (2) the suitability of technologies and their integration in an architecture able to provide the level of flexibility and dynamicity required by current learning environments; and (3) the scalability and performance of the architectures. With challenges in mind, this paper proposes an MEC-enabled architecture that considers SDN/NFV to reconfigure the software and hardware resources of classroom devices in real-time and on-demand. A case study inspired by authentic learning analytics applications extracted from the literature has been proposed to highlight the limitation of the existing solution and demonstrate the added value of our architecture. The experimental results demonstrate acceptable computational performance and efficiency when typical classroom devices such as servers, personal computers or laptops implementing practical learning tools are deployed and reconfigured. Specifically, we investigated experiments with different MEC Apps such as face detector, ASR and physics simulation, each one with different computational requirements. The results point out the potential of our architecture to manage heterogeneous classroom devices in an automatic and efficient way.

As future work, we plan to implement and deploy the proposed architecture in a realistic smart classroom scenario to demonstrate its usefulness with real students. In this sense, we will integrate our architecture in existing platforms able to deploy, dismantle and control the life-cycle of VMs and containers such as OpenStack, as well as control the network infrastructure and the communications of the smart classroom by using OpenDaylight as SDN Controller.

Author Contributions: Conceptualization, A.H.C., J.A.R.-V., F.J.G.C. and M.J.R.-T.; Funding acquisition, G.M.P.; Methodology, A.H.C., J.A.R.-V. and F.J.G.C.; Resources, G.M.P.; Software, F.J.G.C.; Supervision, J.A.R.-V.; Visualization, A.H.C. and F.J.G.C.; Writing—original draft, A.H.C., J.A.R.-V., F.J.G.C., M.J.R.-T. and S.K.S.;

Writing—review and editing, A.H.C., J.A.R.-V., F.J.G.C., M.J.R.-T., S.K.S. and G.M.P. All authors have read and agreed to the published version of the manuscript.

Acknowledgments: This work has been partially supported by the Government of Ireland post-doc fellowship (grant code GOIPD/2018/466 of the Irish Research Council), the Spanish Ministry of Economy and Competitiveness through the Juan de la Cierva Formación program (FJCI-2017-34926), and the European Union via the European Regional Development Fund and in the context of CEITER (Grant agreements No.669074).

Conflicts of Interest: The authors declare no conflict of interest.

References

1. Martín-Gutiérrez, J.; Mora, C.E.; Añorbe-Díaz, B.; González-Marrero, A. Virtual technologies trends in education. *EURASIA J. Math. Sci. Technol. Educ.* **2017**, *13*, 469–486.

2. Timms, M.J. Letting artificial intelligence in education out of the box: Educational cobots and smart classrooms. *Int. J. Artif. Intell. Educ.* **2016**, *26*, 701–712. [CrossRef]

3. Borthwick, A.C.; Anderson, C.L.; Finsness, E.S.; Foulger, T.S. Special article personal wearable technologies in education: Value or villain? *J. Digit. Learn. Teach. Educ.* **2015**, *31*, 85–92. [CrossRef]

4. Ochoa, X.; Worsley, M. Augmenting Learning Analytics with Multimodal Sensory Data. *J. Learn. Anal.* **2016**, *3*, 213–219. [CrossRef]

5. Blikstein, P.; Worsley, M. Multimodal Learning Analytics and Education Data Mining: using computational technologies to measure complex learning tasks. *J. Learn. Anal.* **2016**, *3*, 220–238. [CrossRef]

6. Romano, G.; Schneider, J.; Drachsler, H. Dancing Salsa with Machines—Filling the Gap of Dancing Learning Solutions. *Sensors* **2019**, *19*, 3661, doi:10.3390/s19173661. [CrossRef] [PubMed]

7. Roque, F.; Cechinel, C.; Weber, T.O.; Lemos, R.; Villarroel, R.; Miranda, D.; Munoz, R. Using Depth Cameras to Detect Patterns in Oral Presentations: A Case Study Comparing Two Generations of Computer Engineering Students. *Sensors* **2019**, *19*, 3493, doi:10.3390/s19163493. [CrossRef]

8. Shankar, S.K.; Prieto, L.P.; Rodríguez-Triana, M.J.; Ruiz-Calleja, A. A review of multimodal learning analytics architectures. In Proceedings of the 2018 IEEE 18th International Conference on Advanced Learning Technologies (ICALT), Mumbai, India , 9–13 July 2018; pp. 212–214.

9. Hernández-García, Á.; Conde, M.Á. Dealing with complexity: Educational data and tools for learning analytics. In Proceedings of the Second International Conference on Technological Ecosystems for Enhancing Multiculturality, Salamanca, Spain, 1–3 October 2014; pp. 263–268.

10. Di Mitri, D.; Schneider, J.; Specht, M.; Drachsler, H. The Big Five: Addressing Recurrent Multimodal Learning Data Challenges. In Proceedings of the 8th International Conference on Learning Analytics and Knowledge, Syndey, Australia, 5–9 March 2018.

11. ETSI NFV ISG. *Network Functions Virtualisation (NFV); Network Operator Perspectives on NFV Priorities for 5G*; Technical Report; ETSI White Paper; ETSI: Nice, France, 2017.

12. Singh, S.; Jha, R.K. A survey on software defined networking: Architecture for next generation network. *J. Netw. Syst. Manag.* **2017**, *25*, 321–374, doi:10.1007/s10922-016-9393-9. [CrossRef]

13. Hwang, G.J. Definition, framework and research issues of smart learning environments-a context-aware ubiquitous learning perspective. *Smart Learn. Environ.* **2014**, *1*, 4. [CrossRef]

14. Bautista, G.; Borges, F. Smart classrooms: Innovation in formal learning spaces to transform learning experiences. *Bull. IEEE Tech. Committee Learn. Technol.* **2013**, *15*, 18–21.

15. Muhamad, W.; Kurniawan, N.B.; Yazid, S. Smart campus features, technologies, and applications: A systematic literature review. In Proceedings of the 2017 International Conference on Information Technology Systems and Innovation (ICITSI), Bandung, Indonesia, 23–24 October 2017; pp. 384–391.

16. Xie, W.; Shi, Y.; Xu, G.; Xie, D. Smart classroom-an intelligent environment for tele-education. In Proceedings of the Pacific-Rim Conference on Multimedia, Beijing, China, 24–26 October 2001; Springer: Berlin/Heidelberg, Germany, 2001; pp. 662–668.

17. Snow, C.; Pullen, J.M.; McAndrews, P. Network EducationWare: An open-source web-based system for synchronous distance education. *IEEE Trans. Educ.* **2005**, *48*, 705–712. [CrossRef]

18. Qin, W.; Suo, Y.; Shi, Y. Camps: A middleware for providing context-aware services for smart space. In Proceedings of the International Conference on Grid and Pervasive Computing, Taichung, Taiwan, 3–5 May 2006; Springer: Berlin/Heidelberg, Germany, 2006; pp. 644–653.

19. Suo, Y.; Miyata, N.; Morikawa, H.; Ishida, T.; Shi, Y. Open smart classroom: Extensible and scalable learning system in smart space using web service technology. *IEEE Trans. Knowl. Data Eng.* **2008**, *21*, 814–828. [CrossRef]

20. Muñoz-Cristóbal, J.A.; Rodríguez-Triana, M.J.; Gallego-Lema, V.; Arribas-Cubero, H.F.; Asensio-Pérez, J.I.; Martínez-Monés, A. Monitoring for awareness and reflection in ubiquitous learning environments. *Int. J. Hum.–Comput. Interact.* **2018**, *34*, 146–165. [CrossRef]

21. Serrano-Iglesias, S.; Bote-Lorenzo, M.L.; Gómez-Sánchez, E.; Asensio-Pérez, J.I.; Vega-Gorgojo, G. Towards the enactment of learning situations connecting formal and non-formal learning in SLEs. In *Foundations and Trends in Smart Learning*; Springer: Singapore, 2019; pp. 187–190.

22. Huang, L.S.; Su, J.Y.; Pao, T.L. A context aware smart classroom architecture for smart campuses. *Appl. Sci.* **2019**, *9*, 1837. [CrossRef]

23. Lu, Y.; Zhang, S.; Zhang, Z.; Xiao, W.; Yu, S. A Framework for Learning Analytics Using Commodity Wearable Devices. *Sensors* **2017**, *17*, 1382, doi:10.3390/s17061382. [CrossRef] [PubMed]

24. Miller, H.G.; Mork, P. From Data to Decisions: A Value Chain for Big Data. *IT Profess.* **2013**, *15*, 57–59. [CrossRef]

25. Perales, M.; Pedraza, L.; Moreno-Ger, P. Work-In-Progress: Improving Online Higher Education with Virtual and Remote Labs. In Proceedings of the 2019 IEEE Global Engineering Education Conference (EDUCON), Dubai, UAE, 8–11 April 2019; pp. 1136–1139.

26. Dziabenko, O.; Orduña, P.; García-Zubia, J.; Angulo, I. Remote Laboratory in Education: WebLab-Deusto Practice. In Proceedings of the E-Learn: World Conference on E-Learning in Corporate, Government, Healthcare, and Higher Education, Montréal, QC, Canada, 9–12 October 2012; pp. 1445–1454.

27. University of Deusto and DeustoTech. WebLab-Deusto. 2018. Available online: http://weblab.deusto.es/website (accessed on 1 May 2020).

28. Huertas Celdrán, A.; Garcia, F.; Saenz, J.; De La Torre, L.; Salzmann, C.; Gillet, D. Self-Organized Laboratories for Smart Campus. *IEEE Trans. Learn. Technol.* **2019**. [CrossRef]

29. De La Torre, L.; Neustock, L.T.; Herring, G.; Chacon, J.; Garcia, F.; Hesselink, L. Automatic Generation and Easy Deployment of Digitized Laboratories. *IEEE Trans. Ind. Inform.* **2020**. [CrossRef]

30. Salzmann, C.; Govaerts, S.; Halimi, W.; Gillet, D. The Smart Device specification for remote labs. In Proceedings of the 2015 12th International Conference on Remote Engineering and Virtual Instrumentation (REV), Bangkok, Thailand, 25–27 February 2015; pp. 199–208. [CrossRef]

31. Salzmann, C.; Gillet, D. Smart device paradigm, Standardization for online labs. In Proceedings of the 2013 IEEE Global Engineering Education Conference (EDUCON), Berlin, Germany, 13–15 March 2013; pp. 1217–1221. [CrossRef]

32. Halimi, W.; Salzmann, C.; Jamkojian, H.; Gillet, D. Enabling the Automatic Generation of User Interfaces for Remote Laboratories. In *Online Engineering & Internet of Things*; Springer: Cham, Switzerland, 2018; pp. 778–793._73. [CrossRef]

33. Huertas Celdrán, A.; Gil Pérez, M.; García Clemente, F.J.; Martínez Pérez, G. Automatic monitoring management for 5G mobile networks. *Procedia Comput. Sci.* **2017**, *110*, 328–335. [CrossRef]

34. Salahuddin, M.A.; Al-Fuqaha, A.; Guizani, M.; Shuaib, K.; Sallabi, F. Softwarization of Internet of Things Infrastructure for Secure and Smart Healthcare. *Computer* **2017**, *50*, 74–79. [CrossRef]

35. Muñoz, R.; Nadal, L.; Casellas, R.; Moreolo, M.S.; Vilalta, R.; Fabrega, J.M.; Martinez, R.; Mayoral, A.; Vilchez, F.J. The ADRENALINE testbed: An SDN/NFV packet/optical transport network and edge/core cloud platform for end-to-end 5G and IoT services. In Proceedings of the 2017 European Conference on Networks and Communications (EuCNC), Oulu, Finland, 12–15 June 2017; pp. 1–5. [CrossRef]

36. Nguyen, V.G.; Brunstrom, A.; Grinnemo, K.J.; Taheri, J. SDN/NFV-Based Mobile Packet Core Network Architectures: A Survey. *IEEE Commun. Surv. Tutor.* **2017**, *19*, 1567–1602. [CrossRef]

37. Ge, X.; Zhou, R.; Li, Q. 5G NFV-Based Tactile Internet for Mission-Critical IoT Services. *IEEE Internet Things J.* **2019**. [CrossRef]

38. Qu, K.; Zhuang, W.; Ye, Q.; Shen, X.; Li, X.; Rao, J. Dynamic Flow Migration for Embedded Services in SDN/NFV-Enabled 5G Core Networks. *IEEE Trans. Commun.* **2020**, *68*, 2394–2408. [CrossRef]

39. Huertas Celdrán, A.; Gil Pérez, M.; García Clemente, F.J.; Martínez Pérez, G. Sustainable securing of Medical Cyber-Physical Systems for the healthcare of the future. *Sustain. Comput. Inform. Syst.* **2018**, *19*, 138–146. doi:10.1016/j.suscom.2018.02.010. [CrossRef]

40. Molina Zarca, A.; Bernabe, J.B.; Trapero, R.; Rivera, D.; Villalobos, J.; Skarmeta, A.; Bianchi, S.; Zafeiropoulos, A.; Gouvas, P. Security Management Architecture for NFV/SDN-Aware IoT Systems. *IEEE Internet Things J.* **2019**, *6*, 8005–8020. [CrossRef]

41. Long, Y.; Aleven, V. Educational game and intelligent tutoring system: A classroom study and comparative design analysis. *ACM Trans. Comput.-Hum. Interact. (TOCHI)* **2017**, *24*, 1–27. [CrossRef]

42. Kangas, M.; Koskinen, A.; Krokfors, L. A qualitative literature review of educational games in the classroom: the teacher's pedagogical activities. *Teach. Teach.* **2017**, *23*, 451–470. [CrossRef]

43. Tissenbaum, M.; Slotta, J. Supporting classroom orchestration with real-time feedback: A role for teacher dashboards and real-time agents. *Int. J. Comput.-Support. Collab. Learn.* **2019**, *14*, 325–351. [CrossRef]

44. Holstein, K.; McLaren, B.M.; Aleven, V. Intelligent tutors as teachers' aides: Exploring teacher needs for real-time analytics in blended classrooms. In Proceedings of the Seventh International Learning Analytics & Knowledge Conference, Vancouver, BC, Cannada, 13–17 March 2017; pp. 257–266.

45. Holstein, K.; Hong, G.; Tegene, M.; McLaren, B.M.; Aleven, V. The classroom as a dashboard: Co-designing wearable cognitive augmentation for K-12 teachers. In Proceedings of the 8th International Conference on Learning Analytics and Knowledge, Sydney, Australia, 7–9 March 2018; pp. 79–88.

46. UNESCO Bangkok Office. *School and Teaching Practices for Twenty-First Century Challenges: Lessons from the Asia-Pacific Region—Regional Synthesis Report*; Technical Report; UNESCO: Bangkok, Thailand, 2016; Available online: https://unesdoc.unesco.org/ark:/48223/pf0000244022 (accessed on 1 May 2020).

47. Laal, M.; Laal, M.; Kermanshahi, Z.K. 21st century learning; learning in collaboration. *Procedia-Soc. Behav. Sci.* **2012**, *47*, 1696–1701. [CrossRef]

48. Martinez-Maldonado, R.; Kay, J.; Buckingham Shum, S.; Yacef, K. Collocated collaboration analytics: Principles and dilemmas for mining multimodal interaction data. *Hum.–Comput. Interact.* **2019**, *34*, 1–50. [CrossRef]

49. Praharaj, S.; Scheffel, M.; Drachsler, H.; Specht, M. Multimodal analytics for real-time feedback in co-located collaboration. In Proceedings of the European Conference on Technology Enhanced Learning, Leeds, UK, 3–6 September 2018; pp. 187–201.

50. Schneider, B.; Wallace, J.; Blikstein, P.; Pea, R. Preparing for future learning with a tangible user interface: the case of neuroscience. *IEEE Trans. Learn. Technol.* **2013**, *6*, 117–129. [CrossRef]

51. Maldonado, R.M.; Kay, J.; Yacef, K.; Schwendimann, B. An interactive teacher's dashboard for monitoring groups in a multi-tabletop learning environment. In Proceedings of the International Conference on Intelligent Tutoring Systems, Chania, Greece, 14–18 June 2012; pp. 482–492.

52. Novak, J.D.; Cañas, A.J. *The Theory Underlying Concept Maps and How to Construct and Use Them*; Technical Report; Florida Institute for Human and Machine Cognition: Pensacola, FL, USA, 2008; Available online: http://citeseerx.ist.psu.edu/viewdoc/download?doi=10.1.1.100.8995&rep=rep1&type=pdf (accessed on 1 May 2020).

53. Fleck, R.; Rogers, Y.; Yuill, N.; Marshall, P.; Carr, A.; Rick, J.; Bonnett, V. Actions speak loudly with words: Unpacking collaboration around the table. In Proceedings of the ACM International Conference on Interactive Tabletops and Surfaces, Banff, Canada, 23–25 November 2009; pp. 189–196.

54. Kokotsaki, D.; Menzies, V.; Wiggins, A. Project-based learning: A review of the literature. *Improv. Schools* **2016**, *19*, 267–277. [CrossRef]

55. Topalli, D.; Cagiltay, N.E. Improving programming skills in engineering education through problem-based game projects with Scratch. *Comput. Educ.* **2018**, *120*, 64–74. [CrossRef]

56. Marques, M.; Ochoa, S.F.; Bastarrica, M.C.; Gutierrez, F.J. Enhancing the student learning experience in software engineering project courses. *IEEE Trans. Educ.* **2017**, *61*, 63–73. [CrossRef]

57. Martinez-Maldonado, R. "I Spent More Time with that Team" Making Spatial Pedagogy Visible Using Positioning Sensors. In Proceedings of the 9th International Conference on Learning Analytics & Knowledge, Tempe, AZ, USA, 4–8 March 2019; pp. 21–25.

58. Spikol, D.; Ruffaldi, E.; Cukurova, M. Using multimodal learning analytics to identify aspects of collaboration in project-based learning. In Proceedings of the CSCL'17: The 12th International Conference on Computer Supported Collaborative Learning, Philadelphia, PA, USA, 18–22 June 2017. [CrossRef]

59. Blikstein, P. Using learning analytics to assess students' behavior in open-ended programming tasks. In Proceedings of the 1st International Conference on Learning Analytics and Knowledge, Banff, AB, Canada, 27 February–1 March 2011; pp. 110–116.

60. Ahonen, L.; Cowley, B.U.; Hellas, A.; Puolamäki, K. Biosignals reflect pair-dynamics in collaborative work: EDA and ECG study of pair-programming in a classroom environment. *Sci. Rep.* **2018**, *8*, 1–16. [CrossRef]

61. Goldman, M.; Little, G.; Miller, R.C. Collabode: Collaborative coding in the browser. In Proceedings of the 4th International Workshop on Cooperative And Human Aspects of Software Engineering, Waikiki, HI, USA, 21 May 2011; pp. 65–68.

62. Prinsloo, P.; Slade, S. An elephant in the learning analytics room: The obligation to act. In Proceedings of the Seventh International Learning Analytics & Knowledge Conference, Vancouver, BC, Canada, 13–17 March 2017; pp. 46–55.

63. Shankar, S.K.; Rodríguez-Triana, M.J.; Ruiz-Calleja, A.; Prieto, L.P.; Chejara, P.; Martínez-Monés, A. Multimodal Data Value Chain (M-DVC): A Conceptual Tool to Support the Development of Multimodal Learning Analytics Solutions. *IEEE Rev. Iberoam. Tecnol. Aprendiz.* **2020**, *15*, 113–122. [CrossRef]

Article

Teaching and Learning IoT Cybersecurity and Vulnerability Assessment with Shodan through Practical Use Cases

Tiago M. Fernández-Caramés [1,2,*] and Paula Fraga-Lamas [1,2,*]

1   Department of Computer Engineering, Faculty of Computer Science, Universidade da Coruña, 15071 A Coruña, Spain
2   Centro de investigación CITIC, Universidade da Coruña, 15071 A Coruña, Spain
*   Correspondence: tiago.fernandez@udc.es (T.M.F.-C.); paula.fraga@udc.es (P.F.-L.). Tel.: +34-981-167-000 (ext. 6051) (P.F.-L.)

Received: 28 April 2020; Accepted: 19 May 2020; Published: 27 May 2020

Abstract: Shodan is a search engine for exploring the Internet and thus finding connected devices. Its main use is to provide a tool for cybersecurity researchers and developers to detect vulnerable Internet-connected devices without scanning them directly. Due to its features, Shodan can be used for performing cybersecurity audits on Internet of Things (IoT) systems and devices used in applications that require to be connected to the Internet. The tool allows for detecting IoT device vulnerabilities that are related to two common cybersecurity problems in IoT: the implementation of weak security mechanisms and the lack of a proper security configuration. To tackle these issues, this article describes how Shodan can be used to perform audits and thus detect potential IoT-device vulnerabilities. For such a purpose, a use case-based methodology is proposed to teach students and users to carry out such audits and then make more secure the detected exploitable IoT devices. Moreover, this work details how to automate IoT-device vulnerability assessments through Shodan scripts. Thus, this article provides an introductory practical guide to IoT cybersecurity assessment and exploitation with Shodan.

Keywords: IoT; cybersecurity; Shodan; teaching methodology; use case based learning; security audit; vulnerabilities; cyber-attacks; vulnerability assessment

---

1. Introduction

The Internet of Things (IoT) is a paradigm that involves the connection to the Internet of daily objects, giving remote users and other devices the possibility of monitoring and interacting with them. According to some reports, 75 billion IoT devices will be deployed by 2025 [1] for multiple areas like smart appliances [2], smart agriculture [3], smart healthcare [4,5], or smart cites [6] (a summary of the most relevant IoT application areas is shown in Figure 1). Part of such areas are considered as critical, so their security is key to avoid potential damage.

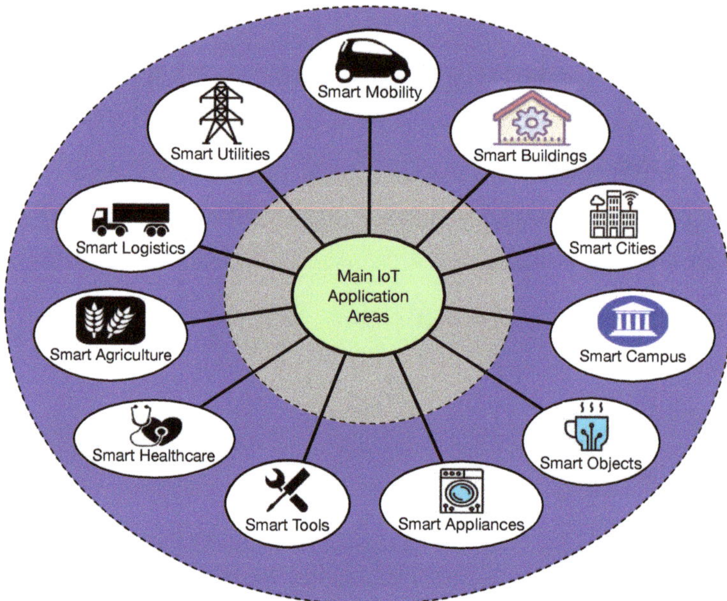

Figure 1. Main Internet of Things (IoT) application areas.

Cybersecurity is a necessary requirement that has to be addressed during the design, implementation and deployment of IoT devices [7,8]. One of the most challenging problems of current IoT devices is that many of them are battery dependent and can be considered as resource-constrained in terms of computational power and memory, which prevents them from implementing certain security features that are common in traditional computers. For instance, public-key cryptography is essential for providing high security for web browsing [9], email exchanges [10], or for storing medical data [11], but the implementation of cryptosystems like Rivest–Shamir–Adleman (RSA) [12] or Elliptic Curve Cryptography (ECC) [13] may not be possible or inefficient for resource-constrained IoT devices. Moreover, such constrained devices may include bugs in their firmware, which in many cases is not possible or easy to update periodically with code patches.

Weak credential security and the lack of basic authentication measures are also common in IoT devices. For instance, such weaknesses were exploited by Mirai, which created a botnet that obtained the administrative credentials of other IoT devices through brute force. Mirai-infected devices, like webcams, Digital Video Recorders (DVRs), or routers, carried out in September 2016 one of the largest Distributed Denial of Service (DDoS) attacks in history, with hundreds of thousands of devices performing simultaneous requests [14]. In many cases, the mentioned weaknesses are related to the fact that, often, product development does not consider security until the final development stages, as an additional layer, instead of considering it as a design requirement.

Although there are a number of recent results of research projects that deal with IoT cybersecurity [15,16], it is almost neglected in many university degrees that are related to the development of IoT products (e.g., electrical engineering, computer science, and computer engineering), so graduated students do not receive in most cases a dense training on IoT security. Moreover, such a lack is also amplified by the difficulty of evaluating a broad range of real IoT devices, which would provide hands-on experience to the students.

To tackle the aforementioned lack, this article includes the following contributions:

- A practical use case-based teaching methodology is proposed. Such a methodology is based on Shodan [17], an online tool that accelerates significantly the IoT device reconnaissance stage, which is usually the most time and resource consuming stage on a cybersecurity assessment.
- This article also provides an introduction to the basics on IoT cybersecurity for future developers, which can harness Shodan Application Programming Interfaces (APIs) to build tools to automate IoT device vulnerability assessments.
- A theoretical and empirical approach to IoT security is provided to help educators to replicate the teaching results obtained by the authors, which have successfully put them in practice in seminars and master courses since 2018. For such a purpose, multiple practical use cases are provided together with useful guidelines to prevent Shodan-based attacks.

The rest of this article is structured as follows. Section 2 analyzes the most recent and relevant work on cybersecurity and IoT security teaching. Section 3 details the proposed teaching methodology. Section 4 details the basics on IoT cybersecurity, including the most common security concerns and the most popular IoT devices and architectures. In addition, Section 4 indicates the main IoT security attacks and describes the typical IoT audit/attack methodology. Section 5 details the basics on Shodan, and Section 6 suggests multiple use cases to put into practice the proposed teaching methodology. Finally, Section 7 is devoted to the conclusions.

2. Related Work

2.1. Cybersecurity Teaching and Learning

Despite the increasing importance of cybersecurity, it is currently not taught extensively in many universities around the world. Some universities have incorporated cybersecurity topics in their study programs [18], but it is still difficult to determine which core competencies should be imparted and then find experts to teach them [19].

Although most cybersecurity teaching still follows the traditional approach based on lectures and labs, some universities have taken them to the cloud and thus imparted virtual cybersecurity lectures on cloud-based platforms. For instance, the authors of [20] describe their experience when teaching a cybersecurity course across two campuses via a virtual classroom. The authors use the Amazon Web Services (AWS) cloud and remark as its main advantage that students perform the exercises in a contained and secure environment without having to deal with cumbersome tasks to set up and configure cybersecurity tools. A similar approach is detailed in [21], where the concept of Cybersecurity Lab as a Service (CLaaS) is proposed to provide cybersecurity experiments to students that can be anywhere and that only need an Internet connection and a device like a laptop, tablet, or a smartphone to carry out the required tasks of the course.

Commercial software and hardware can be used for recreating real-world scenarios for cybersecurity labs, but some researchers find them limited in different aspects and thus created their own frameworks. For instance, in [22], a cybersecurity framework is proposed to develop hands-on experiments rapidly, making use of two incentive models to engage the participants: a model to encourage engineers to contribute with data and experiments and a model to encourage universities to use the contributed data/experiments for education. A different approach is followed in [23], where researchers from Northumbria University (United Kingdom) propose a low-cost and flexible platform that is used as honeypot and that can be integrated with general purpose networks. Similarly, in [24], a modular testbed for teaching cybersecurity in a simulated industrial environment is presented. By using a flipped classroom methodology, students learn about threats associated with the industrial control system domain, develop an educational game, and exercise their soft skills during multiple public presentations.

Regarding IoT cybersecurity teaching, there are not many well documented success cases in the literature. An example is detailed in [25,26], where a course in secure design is described. Such a course is aimed at teaching students how to make user-centered cybersecure products that communicate threats in a better way and that emphasize key decisions to the user. The course consists of classroom instruction, hands-on labs, and prototyping tasks where the students build a conceptual model of a popular IoT smart home product.

Practical experimentation seems to be essential in IoT cybersecurity learning, as it allows the students to retain the knowledge longer than when only traditional lectures are given [27]. For instance, practical experiments carried out with the hardware platform Proxmark3 are key when teaching [28] and evaluating [29] Radio Frequency Identification (RFID) cybersecurity.

Apart from hands-on assignments, other approaches to cybersecurity training include serious games [30]. Examples range from cybersecurity competitions with penetration testing practices [31], capture the flag games [32–34], online learning platforms [35], red versus blue teams [36], or build-it/break-it/fix-it competitions [37]. In this regard, Hendrix et al. [38] investigate whether serious games can be effective cybersecurity training tools. Although their results are generally positive, the authors remark that the evaluation sample size was small and selected. Moreover, the studied games were designed for a very short-term interaction (to be finished in one session), and those papers that included an evaluation only considered immediate short-term impact. Therefore, although the authors considered the positive early indications, the question of whether serious games are effective at training was difficult to answer conclusively. As a result, they concluded that games could represent specific case studies and facilitate case-based learning approaches.

Finally, it is worth mentioning that the vast majority of the IoT cybersecurity literature is aimed at training/teaching university students, but it is also important to consider younger students, who are progressively being taught to code from a younger age. This is why the authors of [39] analyzed potential security and privacy issues that may arise when teaching children how to program the BBC micro:bit platform, which can be used by kids to build their own IoT devices. Other authors focused on promoting training all age groups and on further engaging female students [40]. In such a paper, the authors emphasize the role of problem solving using the scientific method and experiential learning activities.

In contrast to some of the previously mentioned IoT security initiatives, this article proposes to make use of a tool that can be used remotely by any student with just a device able to run a web browser and an Internet connection. Therefore, there is no need for expensive hardware or cloud infrastructure (in the imparted courses, students with smartphones were able to perform most of the methodological steps as if they were using more powerful computers). In addition, although the proposed methodology was specifically conceived for university students, it can be easily adapted to high school teaching. However, it must be pointed out that the practical use cases described later in Section 6 allow for detecting many real-world exposed IoT devices, including some related to industrial or critical scenarios, which may lead to access voluntarily or involuntarily IoT devices and networks that belong to third parties. Therefore, every student/researcher/teacher should check and follow the respective law of his/her country and, of course, not cause any trouble or damage to the involved IoT systems.

2.2. Shodan for IoT Cybersecurity

There are different web-based search engines for generic vulnerability scanning like Zmap [41] or Censys [42], and other online tools like Thingful [43] that are used for gathering data from connected IoT devices, but Shodan is currently the best suited for learning IoT cybersecurity due to the ease of use of its web and API interfaces.

In the last years, several researchers made use of Shodan to evaluate the security of different IoT devices. For instance, in [44], the authors used Shodan to detect devices like routers, firewalls, or web cameras that made use of default credentials or simple passwords. Similarly, in [45], Shodan was used together with other tools like Masscan and Nmap to detect vulnerable DSL routers, printers and IoT devices affected by the Heartbleed bug. In the case of [46], webcams and connected smart cameras were the ones analyzed: the researchers found thousands of them poorly configured or with no security. Other researchers corroborated such results and concluded that webcams are in general barely protected and can be used for cyberattacks [47]. Even more concerning are the results of the work detailed in [48], where numerous vulnerable medical devices were detected using Shodan.

It is also worth mentioning the survey in [49], which emphasizes the need for hardening IoT device security at the view of the ease of use of Shodan and the existence of tools like ShoVAT [50], which automate vulnerability identification. Nessus [51] can also be used for vulnerability identification together with Shodan [52]. Such an assessment can also be carried out through scripts, like the authors of [53] did back in 2014 to detect thousands of exposed webcams, printers, and even traffic control systems. Finally, it must be noted that IoT security analyses can be restricted to certain physical locations or organizations. For instance, in [54], the authors scanned IoT vulnerabilities in Jordan, finding numerous open webcams, industrial control systems and automated tank gauges.

3. Teaching Methodology

This article proposes to structure the learning/teaching process into four main parts:

- Introduction to the main IoT cybersecurity concepts. In this first part, the basics on IoT topics like IoT communications architectures, common IoT devices, and attacks to IoT systems are addressed.
- Introduction to the vulnerability assessment tool. This second part deals with the basics on the use of Shodan.
- Practical use case-based analysis. A set of use cases is given to the students in order to apply to them the proposed analysis methodology. At this point no knowledge of computer programming is required, only a web browser with access to Shodan.
- IoT audit/attack automation. In this final part the students learn how to develop scripts to automate the cybersecurity assessments that in the previous part they performed manually through the Shodan web interface.

The first three of the previous four parts can be carried out by most students that have a minimum knowledge of computers and IoT. Nonetheless, the methodology obtains better results with computer science and electrical engineering students, who usually have a good previous knowledge on how IoT devices and architectures work.

The previously mentioned structured content is typically imparted in an intensive six-week course. Each week, one and a half hours are dedicated to theoretical lectures and another one and a half hours to practical labs. In addition, the students carry out a guided final project on the security of a specific device or field. Although the students choose freely the theme of the project, they are guided by the course instructor to make the most out of the learning experience.

It is important to note that the proposed teaching structure is not lineal throughout the course: most of the theoretical concepts are given during the first three weeks, whereas the last three weeks are essentially focused on the labs and on the final project. Thus, the last three weeks are taught in a flipped classroom format [55], where students are given additional content (e.g., links to IoT security presentations from conferences like DEF CON [56], BlackHat [57], or CCC [58]) that are later discussed during the face-to-face time.

At the end of the course, the students deliver three reports and the corresponding software for the labs and for the final project. The grades are given as follows: 40% of the grade is related to an exam on the theory, 30% is for the lab reports, and 30% is for the final project.

The following syllabus was proposed during the imparted courses:

1. Essential IoT cybersecurity Part I (theory, week 1).

   - Introduction to IoT.
   - Traditional IoT architectures.

- Advanced IoT architectures.

2. Shodan basics (lab 1, week 1).

   - Introduction to Shodan.
   - How Shodan works internally.
   - Shodan basic use.
   - A first search with Shodan.

3. Essential IoT cybersecurity Part II (theory, week 2).

   - Popular IoT devices.
   - Main components of an IoT device.
   - Main IoT-device security problems.

4. Practical IoT security analysis with Shodan (lab 2, week 2).

   - Analysis methodology.
   - Practical use cases.

     – Webcams.
     – Home automation systems.
     – Home devices.

5. Essential IoT cybersecurity Part III (theory, week 3).

   - Common IoT-device vulnerabilities and attacks.

6. Shodan query automation (lab 3, week 3).
7. Final project (weeks 4-6).

It is important to note that teachers should emphasize throughout the lectures the importance of the legal dimension and possible consequences of putting Shodan and similar cybersecurity tools to practice. The next sections of this article provide details on the main topics of the previous syllabus.

4. Essential IoT Cybersecurity

4.1. Main Concerns on IoT Security

As it was previously mentioned in the Introduction of this article, the security of many IoT devices is conditioned by their computational simplicity and their dependence on batteries. The former prevents developers from using security mechanisms that require relevant amounts of computing power or memory, while the latter deters them from implementing complex cryptosystems that can drain the battery fast. There are high-security energy-efficient mechanisms [59], but their implementation is not very common in commercial IoT devices.

Static memory is also a common problem in IoT resource-constrained devices, as software bugs and misbehaviors can be discovered after the deployment stage and thus require to patch the device firmware. Unfortunately, many IoT devices (e.g., sensors and actuators) have not been designed to be updated, like the ones based on Application-Specific Integrated Circuits (ASICs) or whose firmware is stored on a Read-Only Memory (ROM). Other devices are difficult to update for most users, such as the IoT devices that require to disassemble the device and plug a specific hardware programmer in. Nonetheless, it must be mentioned that some IoT devices (usually the most computationally powerful, like smart TVs) can be updated via Over-the-Air (OTA) updates, which allow for receiving periodic firmware patches, dynamic configuration settings, or encryption keys from an IoT provider or a user.

Although most IoT users are essentially concerned by end-device security, IoT networks are composed by other devices like gateways or remote clouds that are also vulnerable to attacks.

As an example, Figure 2 shows, on the right, the main components of a traditional IoT cloud-based architecture, which is currently the most popular among commercial IoT deployments. Such an architecture consists of three layers. The layer at the bottom is the IoT-node layer, which is composed by IoT devices that collect data from their embedded sensors and that receives remote commands from the cloud. IoT nodes connect to the cloud through the gateway layer, which includes local gateways (e.g., wireless or wired access points) and gateways deployed by Internet-Service Provider (ISPs) to reach the Internet. Finally, at the top of the architecture is the cloud, which stores, processes, and provides access to the collected data and allows for sending commands to the IoT devices.

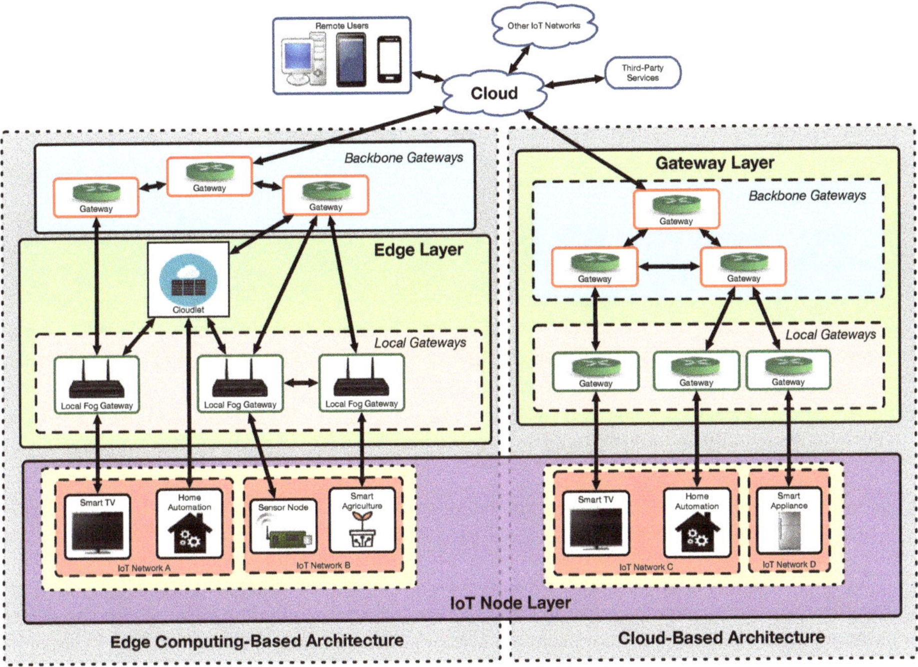

Figure 2. Components of cloud-based and edge computing-based IoT architectures.

4.2. Traditional and Advanced IoT Architectures

Although cloud-based architectures are currently the most popular, they are related to certain security problems that can be prevented by using other advanced architectures. For instance, one of the problems of cloud-based architectures is that they concentrate most of the complex processing and storage on the cloud. This means that the cloud becomes a point-of-failure and, if it has a fault (e.g., due to a cyberattack, to periodic maintenance, or to a power outage), then the whole IoT system stops working properly. Moreover, when a lot of devices perform requests simultaneously, the cloud becomes a bottleneck that slows down the operation of the IoT network due to the excessive workload.

To tackle the previously mentioned issues, decentralized architectures based on edge computing are useful. Figure 2 shows, on the left, the main components of an edge computing based architecture, where three main layers can be distinguished: the IoT node layer, the cloud, and the edge computing layer. The IoT node layer and the cloud operate in a similar way to a traditional cloud-based architecture. The key layer is the edge layer, which provides edge computing services through fog computing gateways and/or cloudlets [60–62]. Fog computing gateways are often devices like Single-Board Computers (SBCs) that provide fast responses and some processing power to the IoT devices in order to reduce latency and the amount of network traffic that is forwarded to the cloud.

Cloudlets have similar objectives, but they are usually high-end computers that perform computing intensive tasks locally. It is also important to note that edge computing nodes can communicate with each other, thus being able to collaborate among them to carry out specific tasks.

There are also other alternative architectures for deploying IoT systems, like the ones based on mist computing [63] or on blockchain [64], which are currently still being studied by industry and academia even for future post-quantum scenarios [65].

4.3. Popular IoT Devices and Cyberattacks

There are many traditional devices that have been enhanced by enabling new features by adding an Internet connection. This is case of TV sets, set-top boxes, home automation systems, intelligent light bulbs, or smart power outlets. Most of them make use of a cloud-based architecture that centralizes request processing in a remote cloud. In this way, if, for instance, a user wants to switch on an smart power outlet through a smartphone app, the user request is first sent to the remote cloud and then the cloud forwards it to the power outlet. This switch-on process is described step by step in Figure 3, where it can be observed that a number of potential problems can arise when user-to-cloud and power outlet-to-cloud communications security is either weak or neglected. Some examples are:

- An evil twin attack can be performed to create a fake local gateway that is able to route IoT device communications to another remote server.

- DoS or DDoS attacks can be performed on the cloud, thus preventing users from sending commands or receiving information from the IoT devices. Similar results may be achieved by carrying out such Dos/DDoS attacks on the communications gateways, which are usually less powerful and less prepared for supporting cyberattacks.

- Weakly encrypted or plain-text communications can be intercepted through sniffers or Man-in-The-Middle (MiTM) attacks, which can gather data on the user or on certain IoT device activities.

- Insecure IoT systems can also be affected by MiTM attacks that are able to modify commands or IoT device responses so as to change the expected behavior of the system.

The impact of the previously mentioned cyberattacks is not only related to traditional homes, but it is amplified due to the broad application fields where IoT is involved, like the deployments related to healthcare [66], smart cities [67], smart infrastructure [68], smart campuses [69], intelligent transport systems [70], or defense and public safety.

In addition, it is important to note that IoT devices like the smart power outlet included in Figure 3 are composed by three different components: hardware, software, and connectivity. Each of such components can be subject to specific attacks and vulnerabilities:

- Hardware attacks. This kind of attacks is related to vulnerabilities that affect certain hardware parts embedded into an IoT device. Examples of such attacks are:

  - Physical attacks.
  - Battery/power removal.
  - Reverse engineering of the hardware.
  - Denial of Service (DoS) attacks to drain batteries.

- Software attacks. These vulnerabilities are related to software bugs or to certain misbehavior that lead to security problems. For instance, some software attacks of this type are:

  - Software reverse engineering.
  - Software vulnerabilities that have or have not been properly patched.
  - Malicious software injection.
  - Weak cryptographic implementations.

- Connectivity attacks. As connectivity is key for implementing the IoT paradigm, IoT devices are vulnerable to traditional attacks aimed at intercepting the exchanged data or at triggering certain behaviors by impersonating an authorized third party. Thus, some of the most relevant connectivity attacks are:

    – DoS attacks.
    – Jamming and radio interference.
    – IoT node impersonation and Sybil attacks.
    – Man-in-the-Middle attacks.
    – Network protocol attacks.

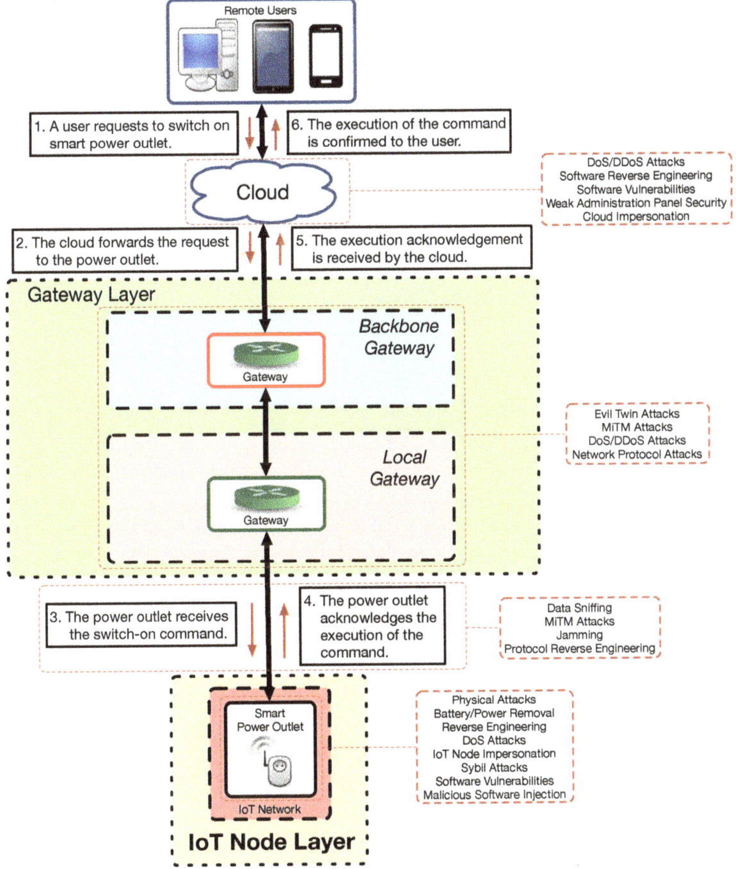

Figure 3. Switching on an IoT-enabled power outlet using a cloud-based architecture.

4.4. IoT Audit/Attack Methodology

Figure 4 illustrates the main steps of the proposed IoT audit/attack methodology, which essentially consists of four phases:

- Reconnaissance. In this phase the auditor/attacker gathers information on the IoT target. The collected data may come from multiple sources (e.g., manufacturers, IoT providers, and hardware datasheets) and includes the traditional port scanning process in order to determine which services are available.

- Audit/Attack plan. The auditor/attacker designs the steps involved in the devised audit/attack strategy and selects the most appropriate tools to implement the plan. In many cases it is necessary to develop specific tools to later exploit certain IoT device vulnerabilities.

- Access to the IoT system. The previously selected tools are used to access the IoT system. Such tools exploit hardware, software, or connectivity vulnerabilities.

- Execution. After accessing the system, an attacker/auditor will put in practice the previously planned strategy to take control of one or more IoT devices. It is common to make use of certain software mechanisms to maintain the access to the IoT system for future intrusions (e.g., by opening a backdoor).

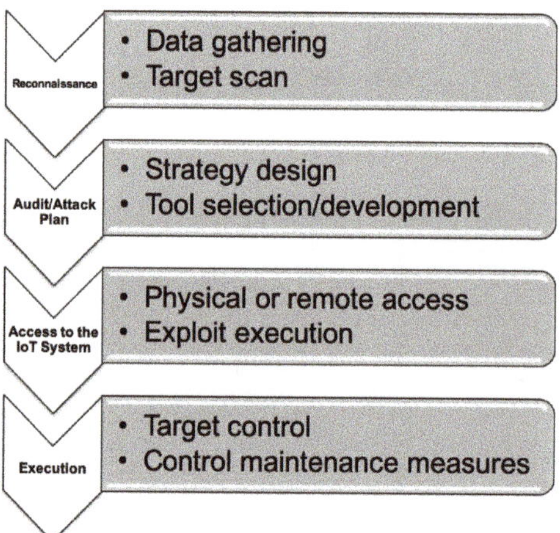

Figure 4. IoT device audit/attack methodology.

Among the previously mentioned phases, the first one (reconnaissance) is usually tedious and requires to dedicate a significant amount of time and resources. However, as it is detailed in the next section, thanks to Shodan, this stage can be noticeably shortened.

5. Shodan Basics

5.1. Aims and Inner Working

Shodan is actually a search engine that scans the Internet IP by IP looking for available services. Such services are detected by parsing banners, which are essentially text that allow for identifying login interfaces or certain service characteristics. An example of a banner is the typical Secure Shell (SSH) login interface, which may provide details on the software of the SSH server or on the computer where it is executed. Shodan indexes banner information and then allows for consulting it through a web interface (shown in Figure 5) and programming APIs.

Underneath, Shodan makes use of crawlers that gather data continuously. There is a crawler network that operates in different countries to prevent IP geo-blocking. Each crawler execute a really simple script that carries out the following steps [71]:

1. A random IPv4 is generated.

2. A random port is selected among the ones supported by Shodan, which are usually related to essential services.

3. The crawler tries to connect to the select IP and port, and if a connection is established, it collects the banner.

4. Go back to step 1.

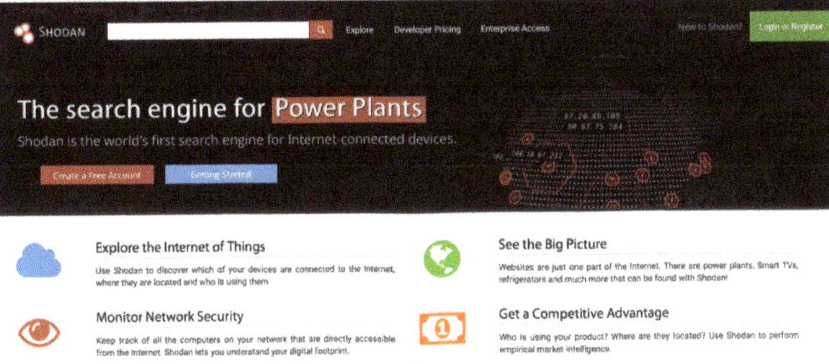

Figure 5. Shodan main web page.

5.2. Basic Use and Web Interface

Shodan can be used as any web search engine, but its use and results differ depending on the user role: there are non-registered users, free registered users, and paid registered users. Each type of user can perform a different number of requests per month, scan a limited number of IPs, and monitor a network with a different maximum of IPs. Differences also exist on the use of certain features, like the application of certain filters or the provided support. Researchers, educators, and students that register with an academic email address can receive a free upgrade (which needs to be requested by email) that enables accessing enough functionality to teach/learn how to use Shodan, but that is usually limited to not to use the accounts to develop commercial applications.

To illustrate the features of Shodan with an example, it can be for instance searched for *"openwrt"*. OpenWrt [72] is actually an operating system based on Linux for embedded devices that can be executed in IoT networks by devices like routers, SBCs, Network Attached Storage (NAS) servers, WiFi extenders, or webcams. The previous Shodan search will lead to a screen like the one shown in Figure 6, which indicates the most relevant sections of the result list page.

Figure 6. Example of Shodan result list page.

When the user clicks on Shodan Maps, the web interface shows a map like the one shown in Figure 7, where the estimated location of the detected OpenWrt devices is depicted. Figure 8 shows the extended information for one of the results obtained in the search. In this screen, on the left, for some devices, detected vulnerabilities are shown. The collected raw data can be accessed by clicking on "View Raw Data".

Figure 7. Shodan Maps interface.

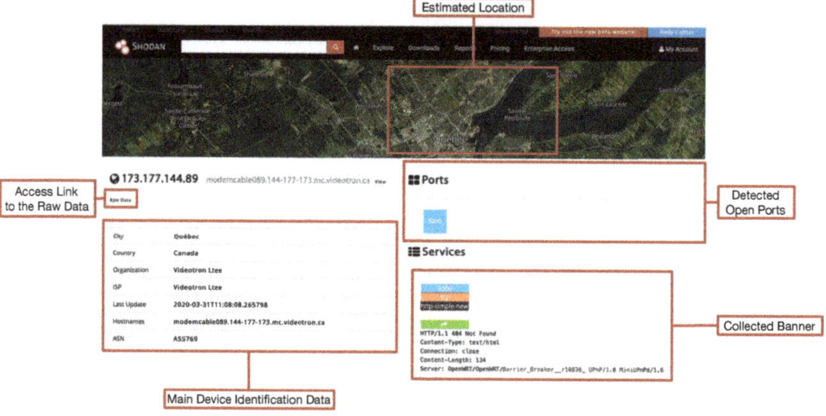

Figure 8. Shodan individual result data page.

Among the multiple features included by Shodan, filters are one of the most useful when looking for specific IoT devices. The following are some of the most relevant:

- country: it specifies the country of the detected devices through an ISO 3166-1 alfa-2 code. For instance, if the previous Shodan search was meant to be limited to the United States, the following query text should be indicated: *"openwrt country:US"*.
- city: it indicates the city of the devices to be located. For instance: *"openwrt city:Barcelona"*.
- geo: it allow for filtering the results depending on their geographical coordinates. If, for instance, the previous results were aimed at obtaining the OpenWrt devices that are located next to Paris city center, the Shodan search would be: *"openwrt geo:48.860151,2.336200"*. Moreover, this filter can received a third parameter that indicates the maximum radius of the search. For example,

the previous search can be modified to obtain the devices that are in a circle of one kilometer around coordinates 48.860151, 2.336200: *"openwrt geo:48.860151,2.336200,1"*.

- net: it filters the results according to an IP range indicated in Classless Inter-Domain Routing (CIDR) notation. An example would be: *"openwrt net:37.13.0.0/16"*.
- port: it allows for filtering the results depending on the detected open ports. For instance, the following Shodan query would return the OpenWrt devices whose port 21 (FTP) is open: *"openwrt port:21"*.
- org: it filters the results according to the organization they belong to. As an example, the following query would indicate the OpenWrt devices that are managed by Amazon: *"openwrt org:amazon"*.

More filters and their parameters can be found in [73].

6. Practical IoT Security Use Case Analysis with Shodan

6.1. Use case Analysis Methodology

6.1.1. Teacher Perspective

From the teaching point of view, the following methodology would be recommended:

- As a first step, the teacher will give the students a list of Shodan searches (like some of the given in Section 6.2).
- Basic analysis. The students analyze the results obtained by each query and determine which IoT device they are looking for and what it is used for. This process usually involves multiple Google searches to look for vendor information like device manuals/datasheets.
- Vulnerability assessment. The students study the vulnerabilities detected by Shodan, they look for default credentials and for other potential cybersecurity problems.

As an example, the previously detailed methodology can be applied to a popular webcam software for Microsoft Windows:

- First, the teacher would give the students the following Shodan query without giving further details on the IoT device: *webcamxp*.
- Next, the students would introduce the query in Shodan and would find out that several thousands of results (more than 5000 as of writing) are shown, most of which are related to a webcam software. As Shodan currently returns a relevant number of honeypots, the students would have to make use of filters to retrieve real webcams. For instance, a refined Shodan search would be: *product:"webcamXP httpd"*.
- After applying the appropriate filters, it is not difficult to find open webcams like the one shown in Figure 9 on the right. It is also straightforward to find further information on the software by looking for *webcamxp manual* through a web search engine.
- Finally, the students will look for security vulnerabilities of the IoT device. In this specific case, the vast majority of the detected webcams neither make use of passwords or implement any kind of access restrictions to control the webcam. The cybersecurity of the hosts that make use of each webcam can be further analyzed with the help of Shodan (e.g., open ports or services), but such a traditional analysis is in general out of the scope of a course focused on IoT cybersecurity.

6.1.2. IoT Researcher Perspective

- Determine the target IoT device.
- Build the Shodan search. This first step requires to determine the most appropriate query and its filters in order to obtain the desired list of target IoT devices.

- Look for additional information on the target IoT device. This process may involve looking for information provided by the manufacturer or for the default credentials indicated in the user manual.

- Vulnerability assessment. In this step it is necessary to analyze the vulnerabilities found by Shodan, the security data provided by the manufacturer or already published Common Vulnerability and Exposure (CVE) reports.

The WebcamXP example given in the previous subsection for the teacher perspective can be used to illustrate how the proposed methodology would be applied by an IoT researcher:

- First, the researcher would set as an objective to find vulnerable webcams that make use of WebcamXP software.

- Next, the researcher will design a first Shodan query (for instance, *webcamxp*) to retrieve the maximum possible amount of IoT devices. Once a webcam is successfully detected (like the one shown in Figure 9 on the right), the Shodan search can be easily refined to avoid collecting data from honeypots and from other devices that include the word *webcamxp* in their banner. For such a purpose, the researcher can analyze the raw information collected by Shodan and select certain fields and values that are highly likely to remain constant for most of the targeted IoT devices. For instance, filtering out by product (Shodan query: *product:"webcamXP httpd"*) or by certain fields of the HTTP header (Shodan query: *"Pragma: no-cache Server: webcamXP"*) can be useful.

- At this point, the researcher may be interested in finding more information on the possibilities that the webcam software can bring to a remote auditor/attacker. For such a purpose, further information on the webcam software is available on the WebcamXP user manual, which can be easily found through a web search engine.

- Although most of the WebcamXP webcams found through Shodan are completely open, the researcher may be interested in exploring further security vulnerabilities of the detected IoT devices. In such a case, CVE repositories like CVE Details allow for searching for WebcamXP vulnerabilities [74], showing three CVE reports: CVE-2008-5862, CVE-2005-1190, and CVE-2005-1189. Shodan academic users can make use of the mentioned CVE IDs and Shodan's vulnerability filter to obtain vulnerable devices directly (Shodan query: *vuln:CVE-2008-5862*).

6.2. Practical use Cases

6.2.1. Webcams and Video Surveillance Systems

Webcams and video surveillance systems probably provide the most common examples on how users lack of knowledge on IoT device security affects privacy and security around the globe: it is currently very easy to find unprotected webcams and video surveillance systems that use their default credentials. Examples of Shodan queries to find this kind of systems are:

- Linksys WVC80N Wireless Internet Camera (Shodan query: *WVC80N*). This is a webcam for home monitoring that is more than 10 years old, but that still is serving in homes and industrial installations. The problem is that many users either use the default credentials (admin/admin) or do not use authentication at all, which causes a privacy problem (an example of screenshot obtained from an open WVC80N webcam is shown in Figure 9 on the left).

- ExacqVision (Shodan query: *"server: wfe"*). This is a video surveillance system that allows for watching and managing multiple webcams through a web interface. The problem is that a significant number of users do not configure authentication or make use of weak/default credentials.

- AXIS webcams (Shodan query: *"port:80 has_screenshot:true"*). As of writing, more than 3000 of these webcams can be found through Shodan, many of them requiring no credentials to watch them.

- AVTECH IP webcams (Shodan query: *linux upnp avtech*). More than 180,000 AVTECH devices can be currently found by Shodan with the previous query, although many of them require credentials to access the video stream. Although the latest firmware versions ask for a verification code, there is a significant number of webcams that make use of the default credentials (admin/admin).

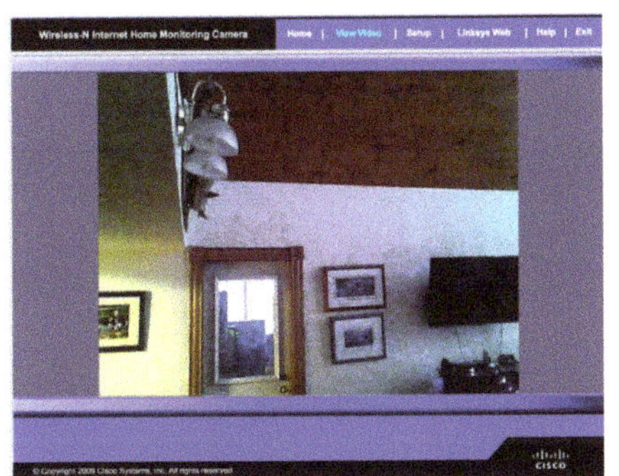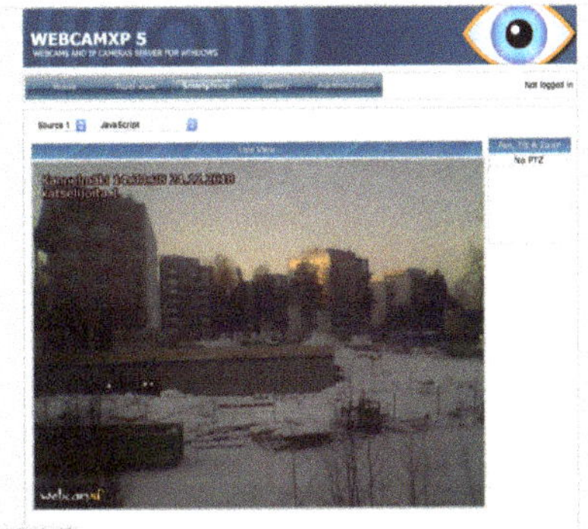

Figure 9. Screenshots of open WVC80N (left) and WebcamXP (right) webcams found with Shodan.

6.2.2. Home Automation Systems

The presence of home automation systems whose security is neglected is also significant. The following are some examples of Shodan queries that will retrieve open or weakly protected home automation system:

- JUNG KNX (Shodan query: *Jung KNX*). This is a home automation system whose smart control panel can be accessed remotely with no need for credentials (an example of such a smart panel is shown in Figure 10 on the left).
- Jeedom (Shodan query: *Jeedom*). It is a French open-source home automation system that usually provides a web interface and, in many cases, an open Message-Queue Telemetry Transport (MQTT) broker.
- Somfy alarm system (Shodan query: *title:"Centrale" Pragma:"no-cache, no-store"*). The previous search allows for locating thousands of Somfy alarm systems, which provide a web interface for remote user authentication.
- Insteon home automation system (Shodan query: *title:"powered by insteon"*). Most of the Insteon installations located through the previous Shodan search require no authentication, so remote users can interact directly with them (a example of an already hacked system is shown in the screenshot in Figure 10 on the right).

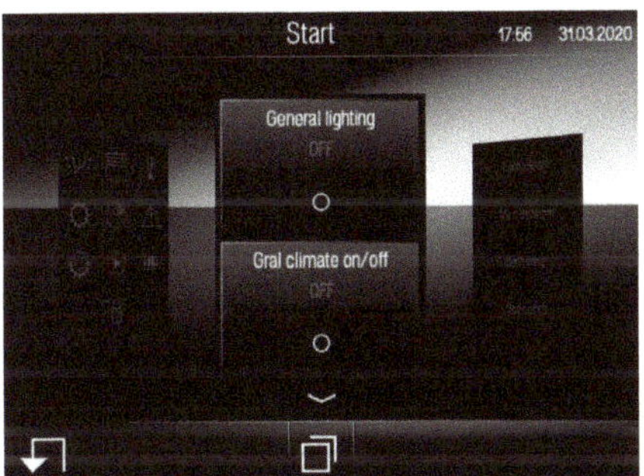

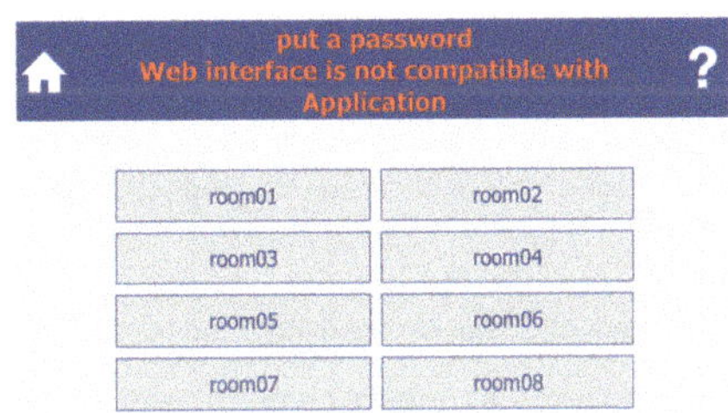

Figure 10. Screenshots of open Jung KNX (**left**) and Insteon (**right**) home automation systems.

- Creston control hub (Shodan query: *Crestron PYNG-HUB*). The web panel of this hub is used by hundreds of users to monitor and control their home automation devices.

6.2.3. Home Devices

Like in the case of home automation systems, many home IoT devices are weakly secured or not secured at all. Some examples of interesting Shodan queries are:

- iKettle (Shodan query: *ikettle*). It is a smart appliance to boil water remotely.
- WebIOPi (Shodan query: *webiopi*). It is a framework for creating and deploying IoT applications with Raspberry Pi. Many installations are not password protected (an screenshot from one of such installations that monitors environmental temperature is shown in Figure 11 on the left).
- Open Virtual Network Computing (VNC) systems (Shodan query: *has_screenshot:true product:VNC "authentication disabled"*). The previous query allows for detecting VNC systems whose authentication has been disabled.
- MQTT brokers (Shodan query: *"MQTT Connection Code: 0" set –alarm*). Although MQTT is very popular among IoT developers, its security, in many cases, is neglected. Thus, the previous Shodan query finds a significant number of open MQTT brokers.
- Yamaha AV receiver (Shodan query: *"HTTP/1.1 406 Not Acceptable" "Server: AV_Receiver"*). Many Yamaha Internet-enabled AV receivers, which provide a remote web interface, have disabled their authentication (a screenshot of one of them is shown in Figure 11 on the right).

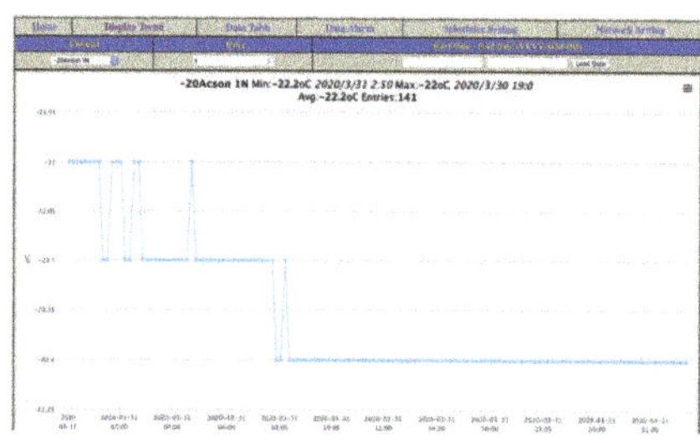

Figure 11. Screenshots of open WebIOPi (**left**) and Yamaha (**right**) installations.

6.3. Automating Attacks

6.3.1. Shodan APIs

Shodan web interface provides a fast way to perform general evaluations on IoT devices or on very specific use cases. However, to automate use case analysis, the APIs provided by Shodan are more appropriate.

Currently, Shodan provides APIs for Python, Ruby, PHP, C#, Go, Haskell, Java, Node.js, Perl, PowerShell, and Rust. Two specific APIs are defined: a Representational State Transfer (REST) API and a streaming API. The REST API is aimed at interacting with Shodan through GET, POST, DELETE, and PUT requests. The streaming API is able to exchange data that is embedded into JavaScript Object Notation (JSON) files. Shodan also provides an additional REST API for exploits [75], which normalizes exploit information after collecting it from multiple vulnerability data sources.

6.3.2. Teaching Shodan Scripting

In order to learn how to automate the manual steps described in Section 6.1, the following tasks can be performed by students/learners:

1. Install the code development environment. This usually requires importing Shodan search library.
2. Perform an initial Shodan query through the code to find a specific version of an IoT device.
3. Modify the code in (2) to print the IP and country of every obtained result.
4. Modify the code in (3) to print, for each detected IoT device that has vulnerabilities, the number of detected exploits according to Shodan exploit REST API.

For instance, the following steps would be needed to perform the previous four tasks when using Python:

1. First, it is necessary to install Python and then install the Shodan module with the command *"pip install shodan"*.
2. An example of the script required for carrying out step 2 is shown in Listing 1 (between lines 1 and 15). Note that, in order to execute the script, it is necessary to indicate the Shodan API key of the developer. In the example, the indicated query can be changed to adapt to the user needs.
3. Listing 1 also shows the part of the script to perform step 3 (between lines 1 and 25). It is worth noting that a 1 second delay is needed, as Shodan may limit the number of requests to one per second.
4. Step 4 can be implemented in Python with the code below line 26 of Listing 1, which makes use of the exploit REST API.

Listing 1: Example of Python script to automate Shodan queries.

```python
import shodan
from time import sleep

SHODAN_API_KEY = "[INSERT HERE YOUR API KEY]"
api = shodan.Shodan(SHODAN_API_KEY)

query = 'webcamxp'

try:

    # Step 2 - Search using Shodan API
    results = api.search(query)
    print('Total number of results: {}'.format(results['total']))

    for result in results['matches']:

      # Step 3 - Print IP and country for every obtained result
        print('IP: {}'.format(result['ip_str'])) # The IP for each result is printed
        #print(result['data']) # To print raw data for each result
        host = api.host(result['ip_str'])
        print('- Country: {0}'.format(host.get('country_name', 'n/a')))
        print('')
        sleep(1) # A 1-second delay is necessary to respect Shodan API restrictions


        # Step 4 - For each device IP, vulnerabilities and exploits are listed
        try:
            if str(host.get('vulns')) != 'None':
                print('------------------- Exploit list -------------------')
                for vulnerability in host.get('vulns'):
                    exploits = api.exploits.search(vulnerability)
                    sleep(1)
                    print('Found {0} exploits for vulnerability "{1}" \n'.format(
                        exploits.get('total'), vulnerability))

        except shodan.APIError as erro:
            print(
                'Error during exploit query: "{0}"'.format(query))
            print('Shodan error: {0}'.format(erro))

except shodan.APIError as e:
    print('Error: {}'.format(e))
```

6.4. Practical Teaching Results

During the last years, the previously described methodology was taught at the University of A Coruña to students of the cybersecurity master program. Each student received three individual Shodan queries and had to first apply the methodology described in Section 6.1.1, and then learned to automate such queries through scripts, following the steps indicated in Section 6.3.2.

As a reference, the following paragraphs summarize the results obtained by the students of the 2020 class where 16 students (two women and fourteen men) took part in the course. Most of them were recent graduates from computer science and electrical engineering programs with good coding skills and basic knowledge on cybersecurity, but almost no previous experience on IoT. They also had no previous practical experience with Shodan.

- Sixteen reports were delivered, with an average of 33 pages per report.
- Different Shodan queries were performed to target 16 specific IoT devices.
- On such 16 IoT devices, 675 non-patched vulnerabilities were found related to already published CVEs.
- Roughly 320 IPs and their running services were analyzed making only use of the information provided by Shodan (no additional scanning tools were used).
- Of the 320 analyzed IoT devices, 87 of them required no credentials to access private data or to manage the device. Moreover, 21 of them made use of the default user or administrator credentials. These results indicate that roughly one out of three analyzed IoT devices could be easily accessed by a remote attacker.

As an example, Table 1 summarizes some of the most relevant results obtained by the students. The following are the main conclusions that can be withdrawn from such results:

Table 1. Summary of the most relevant results obtained by the students of class 2020.

IoT Device	Mootool-Based Webcams	Insteon Smart Home Controller	Somfy Alarm System	IoT Proliphix Thermostats	Cannon VB-M600 Network Cameras	Twonky Media Server
#Shodan Results	141	19	17,294	192	51	3846
#Analyzed Devices	20	19	20	20	20	20
#Devices without Authentication	20	15	-	-	9	20
#Devices with Default Credentials	-	-	2	3	4	-
#Devices Affected by CVEs	4	-	-	-	1	-
#Detected CVEs	66	-	-	-	359	-

1. Mootools-based webcams:

    - Shodan query: *("webcam 7" OR "webcamXP") http.component:"mootools" -401*
    - Relevant results:

        - All the 20 analyzed webcams required no credentials to view their content.
        - Seven of the webcams were used as surveillance cameras in industrial scenarios, while 4 of them were aimed at watching road traffic in specific areas. In addition, 5 of the

        cameras were used as home surveillance systems. The other 4 webcams were used for monitoring public spaces.

- Of the 20 analyzed systems, four of them made use of services and software affected by 66 vulnerabilities that were already documented as CVEs.

2. Insteon smart home controller:

- Shodan query: *title:"powered by insteon"*
- Relevant results:

    - Only 19 results were obtained. Most of the IPs were located in Taiwan and were deployed in homes.
    - Of the 19 IoT systems, 15 of them required no credentials to interact with the smart home system.

3. Somfy alarm system:

- Shodan query: *title:"Centrale" Pragma: "no-cache, no-store"*
- Relevant results:

    - Several of the analyzed systems made use of the default credentials, so attackers could access the alarm system and enable or disable it at will.

4. IoT Proliphix thermostats:

- Shodan query: *title:"Status & Control"*
- Relevant results:

    - A relevant number of the studied IoT systems either used the default user or administration credentials, so a remote attacker could easily watch and manipulate the thermostat.

5. Tesla PowerPack system:

- Shodan query: *http.title:"Tesla PowerPack System"*
- Relevant results:

    - Some of the analyzed IoT systems could be accessed as administrator by making use of the default credentials. However, most of the systems found through Shodan were actually classified as honeypots.

6. Cannon VB-M600 network camera system:

- Shodan query: *title:"Network Camera VB-M600" "200 ok server: vb"*
- Relevant results:

    - Of the 20 analyzed systems, nine of them could be accessed with no credentials, while four made of use of the default credentials.
    - The software used by these systems were affected by 359 vulnerabilities documented through already published CVEs. Such vulnerabilities were essentially related to the use of outdated versions of Linux and Apache Tomcat.

7. Twonky media server:

- Shodan query: *"product:TwonkyMedia UPnP" http.title:"Twonky Server"*
- Relevant results:

- All the devices found through the indicated Shodan query were completely open, so remote attackers can access the shared media content.

Given these results, as one of the students indicated in his report, "it can be concluded that Shodan is a really powerful cybersecurity tool that is able to expose IoT device misconfigurations and vulnerabilities in an easy and fast way; the possibility of using Shodan for automatic IoT vulnerability assessments emphasizes the importance of taking care of security during IoT device installation and configuration, and makes it necessary to patch their software periodically".

Finally, it is worth mentioning that, during the course, there were no major problems respect to the use of Shodan. The only relevant issues arose in relation to the following two topics:

- API-based development. During the development of the scripts the students had problems when dealing with the Python wrapper API, as part of it was not properly documented.
- Critical infrastructure vulnerabilities. In case of finding vulnerabilities that affected critical infrastructures, the students were told to warn the instructor so that he/she could take the appropriate measures (e.g., to warn the company/entity through the university on the encountered problems). For instance, during the course, the mentioned procedure was used by a student that found VoIP communications system of a military company that used the default credentials.

6.5. Preventing Shodan-Based Attacks on IoT Devices: Best Practices

The previous sections show that Shodan is a really powerful tool for performing IoT cybersecurity audits and attacks. In the case of the former, an auditor can give the following recommendations to prevent the audited IoT devices from being attacked through Shodan:

- Check your IP or your organization IP range to determine whether your IoT devices are already indexed by Shodan. If they are indexed, verify their connectivity needs, trying to minimize the number of them that accept incoming connections.
- Minimize the number of open ports. In addition, make use of firewalls to prevent potential intrusions.
- Always try to use HTTPS instead of HTTP. This may be difficult to implement in certain resource-constrained IoT devices. In addition, please note that it is very complex to have an individual (no self-signed) certificate for each IoT device, so try to implement additional security layers.
- Whenever possible, try to use a Virtual Private Network (VPN).
- Whenever possible, modify your IoT device banners and the exposed ports to make the reconnaissance stage difficult for potential attackers. For instance, move the necessary ports to a range that is not scanned by Shodan crawlers.
- Block Shodan crawler IPs to prevent IoT devices from being indexed. A good list of such crawler IPs can be found in [76].
- In case the IoT device cannot be protected from being indexed by Shodan:

  - Never use default or really common credentials (e.g., "admin", "1234").
  - Try to use long usernames and passwords to avoid brute-force attacks.
  - Update credentials periodically.
  - Keep IoT device firmware updated.

6.6. Additional Course Topics

This article described the content, structure, and methodology applied to a 6-week course that, due to time restrictions, is focused on detecting vulnerable IoT devices that are publicly exposed on the Internet. However, a complete IoT cybersecurity program should extend the proposed syllabus and address other relevant topics, like:

- Ethical hacking. Students should learn about the implications and differences among black hat, white hat, and gray hat hackers, which can make use of Shodan with different purposes.
- Legality. Cybersecurity researchers and students should be fully aware of the legal dimension and potential consequences of making use of Shodan and other security tools.
- Defense against IoT attacks. Although Section 6.5 enumerates different recommendations to protect IoT devices against Shodan-based attacks, IoT devices are exposed to many more attacks, like the ones indicated in Section 4.3. Therefore, it is necessary to teach students how to protect IoT devices from physical attacks, software/hardware reverse engineering, malicious firmware updates, or rogue wireless access points.
- Critical infrastructure cybersecurity. IoT devices can be deployed in environments whose infrastructure can be considered as strategical or critical due to the impact that cyberattacks can have on them. For instance, cyberattacks on certain industries (e.g., chemical plants and power plants) or infrastructure (e.g., bridges, dams, ports and railways) can have terrible consequences, so students need to be trained on the specific characteristics of such environments and on the most commonly used monitoring devices (e.g., Programmable Logic Controllers (PLCs) and Industrial Control Systems (ICSs)).
- Mobile device security. A mobile device, like a wearable, a smartphone, or a tablet, can be considered as a specific type of IoT device that provides users with certain communications services and monitoring capabilities (e.g., by making use of embedded sensors like accelerometers, gyroscopes, and GPS). For instance, unfortunately, Shodan can find thousands of open Android devices (Shodan query: *port:5555 debug*) that require no credentials for accessing the internal memory, for installing new applications, or for taking pictures with the embedded camera. Therefore, students should understand how the most popular mobile operating systems and devices work, and how they can be protected against cyberattacks.
- Platform security. Robot, cobot, Unmanned Aerial Vehicle (UAVs), or Augmented/Mixed/Virtual (AR/MR/VR) platforms can be considered as IoT platforms that make use of sensors, actuators, and communications subsystems that are expected to suffer from cybersecurity attacks. Students should understand how to keep information protected, defend against unauthorized use, tampering, or even physical damage.

7. Conclusions

IoT cybersecurity is a topic whose importance has been growing in the last years, but that has not been extensively covered in IT university programs. To ease IoT cybersecurity teaching, this article proposed a practical use case-based methodology that relies on Shodan, a search engine for exploring the Internet that is able to find connected IoT devices. Thus, students only need a web browser and Internet connectivity to carry out practical cybersecurity audits and analyses. Multiple practical examples have been given to discover IoT-enabled devices like webcams or home automation systems, which usually make use of default credentials and/or of weak authentication mechanisms. In addition, the article showed examples of scripts that allow for using Shodan to automate IoT-device vulnerability assessments. Thanks to the previous contributions, this article provided teachers and developers the basics for creating future Shodan-based IoT cybersecurity courses and vulnerability assessment software.

Author Contributions: T.M.F.-C. and P.F.-L. contributed equally to the involved analysis and writing. T.M.F.-C. conceived the article and performed the data collection. All authors have read and agreed to the published version of the manuscript.

Sensors **2020**, *20*, 3048

Funding: This work has been funded by the Xunta de Galicia (ED431G2019/01), the Agencia Estatal de Investigación of Spain (TEC2016-75067-C4-1-R, RED2018-102668-T, PID2019-104958RB-C42) and ERDF funds of the EU (AEI/FEDER, UE).

Conflicts of Interest: The author declares no conflicts of interest.

References

1. HIS, Internet of Things (IoT) Connected Devices Installed Base Worldwide from 2015 to 2025 (In Billions). Available online: https://www.statista.com/statistics/471264/iot-number-of-connected-devices-worldwide/ (accessed on 9 April 2020).

2. Blanco-Novoa, O.; Fernández-Caramés, T.M.; Fraga-Lamas, P.; Castedo, L. A Cost-Effective IoT System for Monitoring Indoor Radon Gas Concentration. *Sensors* **2018**, *18*, 2198. [CrossRef] [PubMed]

3. Ayaz, M.; Ammad-Uddin, M.; Sharif, Z.; Mansour, A.; Aggoune, E. M. Internet-of-Things (IoT)-Based Smart Agriculture: Toward Making the Fields Talk. *IEEE Access* **2019**, *7*, 129551–129583. [CrossRef]

4. Fernández-Caramés, T.M.; Froiz-Míguez, I.; Blanco-Novoa, O.; Fraga-Lamas, P. Enabling the Internet of Mobile Crowdsourcing Health Things: A Mobile Fog Computing, Blockchain and IoT Based Continuous Glucose Monitoring System for Diabetes Mellitus Research and Care. *Sensors* **2019**, *19*, 3319.

5. Alam, M. M.; Malik, H.; Khan, M. I.; Pardy, T.; Kuusik, A.; Le Moullec, Y. A Survey on the Roles of Communication Technologies in IoT-Based Personalized Healthcare Applications. *IEEE Access* **2018**, *6*, 36611–36631. [CrossRef]

6. Fraga-Lamas, P.; Celaya-Echarri, M.; Lopez-Iturri, P.; Castedo, L.; Azpilicueta, L.; Aguirre, E.; Suárez-Albela, M.; Falcone, F.; Fernández-Caramés, T.M. Design and Experimental Validation of a LoRaWAN Fog Computing Based Architecture for IoT Enabled Smart Campus Applications. *Sensors* **2019**, *19*, 3287.

7. Lu, Y.; Xu, L.D. Internet of Things (IoT) Cybersecurity Research: A Review of Current Research Topics. *IEEE Int. Things* . **2019**, *6*, 2103–2115.

8. Augusto-Gonzalez, J.; Collen, A.; Evangelatos, S.; Anagnostopoulos, M.; Spathoulas, G.; Giannoutakis, K. M.; Votis, K.; Tzovaras, D.; Genge, B.; Gelenbe, E.; Nijdam, N. A. From internet of threats to internet of things: A cyber security architecture for smart homes. In Proceedings of the 2019 IEEE 24th International Workshop on Computer Aided Modeling and Design of Communication Links and Networks (CAMAD), Limassol, Cyprus, 11–13 September 2019; pp. 1–6.

9. IETF, RFC 8446: The Transport Layer Security (TLS) Protocol Version 1.3. Aug. 2018. Available online: https://tools.ietf.org/html/rfc8446 (accessed on 9 April 2020).

10. IETF, RFC 3156: MIME security with OpenPGP. Aug. 2000. Available online: https://tools.ietf.org/html/rfc3156 (accessed on 9 April 2020).

11. Tseng, C.H.; Wang, S.H.; Tsaur, W.J. Hierarchical and Dynamic Elliptic Curve Cryptosystem Based Self-Certified Public Key Scheme for Medical Data Protection. *IEEE Trans. Reliab.* **2015**, *64*, 1078–1085. [CrossRef]

12. Rivest, R. L.; Shamir, A.; Adleman, L. M. A method for obtaining digital signatures and public-key cryptosystems. *Commun. ACM* **1978**, *21*, 120–126.

13. Koblitz, N. Elliptic curve cryptosystems. *Math. Comput.* **1987**, *48*, 203–209.

14. Kolias, C.; Kambourakis, G.; Stavrou, A.; Voas, J. DDoS in the IoT: Mirai and Other Botnets. *Computer* **2017**, *50*, 80–84. [CrossRef]

15. Ghavami, N.; Volkamer, M.; Haller, P.; Sánchez, A.; Dimas, M. GHOST-Safe-Guarding Home IoT Environments with Personalised Real-Time Risk Control. In *Security in Computer and Information Sciences: First International ISCIS Security Workshop 2018*; Euro-CYBERSEC; Springer: London, UK, 2018.

16. Meneghello, F.; Calore, M.; Zucchetto, D.; Polese, M.; Zanella, A. IoT: Internet of Threats? A Survey of Practical Security Vulnerabilities in Real IoT Devices. *IEEE Int. Things J.* **2019**, *6*, 8182–8201.

17. Shodan Official Web Page. Available online: https://www.shodan.io (accessed on 9 April 2020).

18. Hölbl, M.; Welzer, T. Experience with Teaching Cybersecurity. In Proceedings of the 27th EAEEIE Annual Conference, Grenoble, France, 7–9 June 2017; pp. 1–4.

19. Parekh, G.; DeLatte, D.; Herman, G.L.; Oliva, L.; Phatak, D.; Scheponik, T.; Sherman, A.T. Identifying Core Concepts of Cybersecurity: Results of Two Delphi Processes. *IEEE Trans. Educ.* **2018**, *61*, 11–20. [CrossRef]

20. Salah, K.; Hammoud, M.; Zeadally, S. Teaching Cybersecurity Using the Cloud. *IEEE Trans. Learn. Technol.* **2015**, *8*, 383–392. [CrossRef]

21. Tunc, C.; Hariri, S.; De La Peña Montero, F.; Fargo, F.; Satam, P.; Al-Nashif, Y. Teaching and Training Cybersecurity as a Cloud Service. In Proceedings of the 2015 International Conference on Cloud and Autonomic Computing, Boston, MA, USA, 21–25 September 2015; pp. 302–308.

22. Wang, L.; Tian, Z.; Gu, Z.; Lu, H. Crowdsourcing Approach for Developing Hands-On Experiments in Cybersecurity Education. *IEEE Access* **2019**, *7*, 169066–169072.

23. Eliot, N.; Kendall, D.; Brockway, M. A Flexible Laboratory Environment Supporting Honeypot Deployment for Teaching Real-World Cybersecurity Skills. *IEEE Access* **2018**, *6*, 34884–34895. [CrossRef]

24. Čeleda, P.; Vykopal, J.; Švábenský, V.; Slavíček, K. KYPO4INDUSTRY: A Testbed for Teaching Cybersecurity of Industrial Control Systems. In Proceedings of the 51st ACM Technical Symposium on Computer Science Education, Portland, Oregon, USA, 11–14 March 2020; pp. 1026–1032.

25. Sharevski, F.; Trowbridge, A.; Westbrook, J. Novel approach for cybersecurity workforce development: A course in secure design. In Proceedings of the IEEE Integrated STEM Education Conference (ISEC), Princeton, NJ, USA, 11 March 2018; pp. 175–180.

26. Sharevski, F.; Treebridge, P.; Westbrook, J. Experiential User-Centered Security in a Classroom: Secure Design for IoT. *IEEE Commun. Mag.* **2019**, *57*, 48–53. [CrossRef]

27. Ban, Y.; Okamura, K.; Kaneko, K. Effectiveness of Experiential Learning for Keeping Knowledge Retention in IoT Security Education. In Proceedings of the 6th IIAI International Congress on Advanced Applied Informatics, Hamamatsu, Japan, 9–13 July 2017; 699–704.

28. Figueroa, S.; Carías, J. F.; Añorga, J.; Arrizabalaga, S.; Hernantes, J. A RFID-based IoT Cybersecurity Lab in Telecommunications Engineering. In Proceedings of Technologies Applied to Electronics Teaching Conference, La Laguna, Spain, 20-22 June 2018, pp. 1–8.

29. Fernández-Caramés, T.M.; Fraga-Lamas, P.; Suárez-Albela, M.; Castedo, L. A methodology for evaluating security in commercial RFID systems. In *Radio Frequency Identification*; IntechOpen: London, UK, 2017. [CrossRef]

30. Topham, L.; Kifayat, K.; Younis, Y. A.; Shi, Q.; Askwith, B. Cyber security teaching and learning laboratories: A survey. *Inf. Secur.* **2016**, *35*, 51. [CrossRef]

31. Bock, K.; Hughey, G.; Levin, D. King of the Hill: A Novel Cybersecurity Competition for Teaching Penetration Testing. In Proceedings of the 2018 USENIX Workshop on Advances in Security Education (ASE 18), Baltimore, MD, USA, 13 August 2018; pp. 1–9.

32. DEF CON 27 Capture the Flag. Available online: https://www.defcon.org/html/defcon-27/dc-27-ctf.html (accessed on 9 April 2020).

33. Ford, V.; Siraj, A.; Haynes, A.; Brown, E. Capture the flag unplugged: an offline cyber competition. In Proceedings of the 2017 ACM SIGCSE Technical Symposium on Computer Science Education, Seattle Washington, USA, 8–11 March 2017; pp. 225–230.

34. Chapman, P.; Burket, J.; Brumley, D. PicoCTF: A Game- Based Computer Security Competition for High School Students. In Proceedings of the 2014 USENIX Summit on Gaming, Games, and Gamification in Security Education (3GSE 14), USENIX Association, San Diego, CA, USA, 18 August 2014; pp. 1–10.

35. Root Me. The Fast, Easy, and Affordable Way to Train Your Hacking Skills. Challenge Your Hacking Skills. Available online: https://www.root-me.org/?lang=en (accessed on 9 April 2020).

36. Vykopal, J.; Vizvary, M.; Oslejsek, R.; Celeda, P.; Tovarnak, D. Lessons Learned From Complex Hands-on Defence Exercises in a Cyber Range. In Proceedings of the 2017 IEEE Frontiers in Education Conference (FIE), Indianapolis, IN, USA, 18–21 October 2017; pp. 1–8.

37. Ruef, A.; Hicks, M.; Parker, J.; Levin, D.; Mazurek, M.L.; Mardziel, P. Build it, break it, fix it: Contesting secure development. In Proceedings of the 2016 ACM SIGSAC Conference on Computer and Communications Security, Vienna Austria, 24–28 October 2016; pp. 690–703.

38. Hendrix, M.; Al-Sherbaz, A.; Bloom, V. Game based cyber security training: Are serious games suitable for cyber security training?. *Int. J. Serious Games* **2016**, *3*, 1.

39. Knowles, B.; Finney, J.; Beck, S.; Devine, J. What children's imagined uses of the BBC micro:bit tells us about designing for their IoT privacy, security and safety. In Proceedings of Living in the Internet of Things: Cybersecurity of the IoT, London, UK, 28-29 March 2018 March 2018; pp. 1–6.

40. Liu, X.; Murphy, D. Engaging females in cybersecurity: K through Gray. In Proceedings of the 2016 IEEE Conference on Intelligence and Security Informatics (ISI), Tucson, AZ, USA, 17 November 2016; pp. 255–260.

41. Zmap Official Web Page. Available online: https://zmap.io (accessed on 9 April 2020)

42. Censys Official Web Page. Available online: https://censys.io (accessed on 9 April 2020)

43. Thingful Official Web Page. Available online: https://www.thingful.net (accessed on 9 April 2020)

44. Albataineh, A.; Alsmadi, I. IoT and the Risk of Internet Exposure: Risk Assessment Using Shodan Queries. In Proceedings of the 2019 IEEE 20th International Symposium on "A World of Wireless, Mobile and Multimedia Networks" (WoWMoM), Washington, DC, USA, 10–12 June 2019; pp. 1–5.

45. Markowsky, L.; Markowsky, G. Scanning for vulnerable devices in the Internet of Things. In Proceedings of the 2015 IEEE 8th International Conference on Intelligent Data Acquisition and Advanced Computing Systems: Technology and Applications (IDAACS), Warsaw, Poland, 24–26 September 2015; pp. 463-467.

46. Bugeja, J.; Jönsson, D.; Jacobsson, A. An Investigation of Vulnerabilities in Smart Connected Cameras. In Proceedings of the 2018 IEEE International Conference on Pervasive Computing and Communications Workshops (PerCom Workshops), Athens, Greece, 19 March 2018; pp. 537–542.

47. Vlajic, N.; Zhou, D. IoT as a Land of Opportunity for DDoS Hackers. *Computer* **2018**, *51*, 26–34.

48. McMahon, E.; Williams, R.; El, M.; Samtani, S.; Patton, M.; Chen,H. Assessing medical device vulnerabilities on the Internet of Things. In Proceedings of the 2017 IEEE International Conference on Intelligence and Security Informatics (ISI), Beijing, China, 22–24 July 2017; pp. 176–178.

49. Rae, J.S.; Chowdhury, M.M.; Jochen, M. Internet of Things Device Hardening Using Shodan.io and ShoVAT: A Survey. In Proceedings of the 2019 IEEE International Conference on Electro Information Technology (EIT), Brookings, SD, USA, 20–22 May 2019; pp. 379–385.

50. Genge, B.; Enăchescu, C. ShoVAT: Shodan-based vulnerability assessment tool for Internet-facing services. *Secur. Commun. Networks* **2015**, *9*, 2696–2714. [CrossRef]

51. Nessus Official Web Page. Available online: https://www.tenable.com/products/nessus/nessus-professional (accessed on 9 April 2020)

52. Williams, R.; McMahon, E.; Samtani, S.; Patton, M; Chen, H. Identifying vulnerabilities of consumer Internet of Things (IoT) devices: A scalable approach. In Proceedings of the 2017 IEEE International Conference on Intelligence and Security Informatics (ISI), Beijing, China, 22–24 July 2017; pp. 179–181.

53. Patton, M.; Gross, E.; Chinn, R.; Forbis, S.; Walker, L.; Chen, H. Uninvited Connections: A Study of Vulnerable Devices on the Internet of Things (IoT). In Proceedings of the 2014 IEEE Joint Intelligence and Security Informatics Conference, The Hague, Netherlands, 24–26 September 2014; pp. 232–235.

54. Al-Alami, H.; Hadi, A.; Al-Bahadili, H. Vulnerability scanning of IoT devices in Jordan using Shodan. In Proceedings of the 2017 2nd International Conference on the Applications of Information Technology in Developing Renewable Energy Processes & Systems (IT-DREPS), Amman, Jordan, 6–7 December 2017; pp. 1–6.

55. Mason, G.S.; Shuman, T.R.; Cook, K.E. Comparing the Effectiveness of an Inverted Classroom to a Traditional Classroom in an Upper-Division Engineering Course. *IEEE Trans. Educ.* **2013**, *56*,430–435.

56. DEF CON Conference Official Web Page. Available online: https://www.defcon.org (accessed on 9 April 2020)

57. Black Hat Conference Official Web Page. Available online: https://www.blackhat.com (accessed on 9 April 2020)

58. Chaos Computer Club Official Media Repository. Available online: https://media.ccc.de (accessed on 9 April 2020)

59. Suárez-Albela, M.; Fraga-Lamas, P.; Castedo, L.; Fernández-Caramés, T.M. Clock frequency impact on the performance of high-security cryptographic cipher suites for energy-efficient resource-constrained IoT devices. *Sensors* **2019**, *19*, 3868. [CrossRef] [PubMed]

60. Fraga-Lamas, P.; Lopez-Iturri, P.; Celaya-Echarri, M.; Blanco-Novoa, O.; Azpilicueta, L.; Varela-Barbeito, J.; Falcone, F.; Fernández-Caramés, T. M. Design and Empirical Validation of a Bluetooth 5 Fog Computing Based Industrial CPS Architecture for Intelligent Industry 4.0 Shipyard Workshops. *IEEE Access* **2020**, *8*, 45496–45511. [CrossRef]

61. Perera, C.; Qin, Y.; Estrella, J.C.; Reiff-Marganiec, S.; Vasilakos, A.V. Fog computing for sustainable smart cities: A survey. *ACM Comput. Surv. (CSUR)* **2017**, *50*, 1-43. [CrossRef]

62. Alturki, B.; Reiff-Marganiec, S.; Perera, C.; De, S. Exploring the Effectiveness of Service Decomposition in Fog Computing Architecture for the Internet of Things. *IEEE Trans. Sustain. Comput.* **2019**.

63. Suárez-Albela, M.; Fraga-Lamas, P.; Fernández-Caramés, T.M. A Practical Evaluation on RSA and ECC-Based Cipher Suites for IoT High-Security Energy-Efficient Fog and Mist Computing Devices. *Sensors* **2018**, *18*, 3868. [CrossRef]

64. Fernández-Caramés, T.M.; Fraga-Lamas, P. A Review on the Application of Blockchain for the Next Generation of Cybersecure Industry 4.0 Smart Factories. *IEEE Access* **2019**, *7*, 45201–45218. [CrossRef]

65. Fernández-Caramés, T.M.; Fraga-Lamas, P. Towards post-quantum blockchain: A review on blockchain cryptography resistant to quantum computing attacks. *IEEE Access* **2020**, *8*, 21091–21116. [CrossRef]

66. Strielkina, A.; Illiashenko, O.; Zhydenko, M.; Uzun,D. Cybersecurity of healthcare IoT-based systems: Regulation and case-oriented assessment. In Proceedings of the 2018 IEEE 9th International Conference on Dependable Systems, Services and Technologies (DESSERT), Kiev, Ukraine, 24–27 May 2018; pp. 67–73.

67. Alrashdi, I.; Alqazzaz, A.; Aloufi, E.; Alharthi, R.; Zohdy, M.; Ming, H. AD-IoT: Anomaly Detection of IoT Cyberattacks in Smart City Using Machine Learning. In Proceedings of the 2019 IEEE 9th Annual Computing and Communication Workshop and Conference (CCWC), Las Vegas, NV, USA, 7–9 January 2019; pp. 305–310.

68. Trotter, L.; Harding, M.; Mikusz, M.; Davies, N. IoT-Enabled Highway Maintenance: Understanding Emerging Cybersecurity Threats. *IEEE Pervasive Comput.* **2018**, *17*, 23–34.

69. Webb, J.; Hume, D. Campus IoT collaboration and governance using the NIST cybersecurity framework. In Proceedings of the Living in the Internet of Things: Cybersecurity of the IoT, London, UK, 28–29 March 2018; pp. 1–7.

70. Frötscher, FA.; Monschiebl, B.; Drosou, A.; Gelenbe, E.; Reed, M. J.; Al-Naday, M. Improve cybersecurity of C-ITS Road Side Infrastructure Installations: the SerIoT—Secure and Safe IoT approach. In Proceedings of the 2019 IEEE International Conference on Connected Vehicles and Expo (ICCVE), Graz, Austria, 4–8 November 2019; pp. 1–5.

71. Matherly, J. Complete Guide to Shodan. Collect. Analyze. Visualize. Make Internet Intelligence Work for You. Available online: https://www.amazon.com/Complete-Guide-Shodan-Visualize-Intelligence-ebook/dp/B01CDIU880 (accessed on 9 April 2020).

72. OpenWrt Official Web Page. Available online: https://openwrt.org (accessed on 9 April 2020).

73. Javier Olmedo GitHub Repository for Shodan Filters. Available online: https://github.com/JavierOlmedo/shodan-filters (accessed on 9 April 2020).

74. Vulnerabilities Collected by CVE Details for WebcamXP. Available online: https://www.cvedetails.com/vulnerability-list/vendor_id-2917/Webcamxp.html (accessed on 17 May 2020).

75. Shodan Exploit API. Available online: https://developer.shodan.io/api/exploits/rest (accessed on 9 April 2020).

76. List of Shodan Crawler IPs. Available online: https://wiki.ipfire.org/configuration/firewall/blockshodan (accessed on 9 April 2020).